ICA

XLV. 6. 577

ÆDES CHRISTI
in Academia Oxoniensi

CHAS. W.E. 1904.

BONDING AND STRUCTURE
Structural Principles in
Inorganic and Organic Chemistry

ELLIS HORWOOD SERIES IN INORGANIC CHEMISTRY

Series Editor: J. BURGESS, Department of Chemistry, University of Leicester

Inorganic chemistry is a flourishing discipline in its own right and also plays a key role in many areas of organometallic, physical, biological, and industrial chemistry. This series is developed to reflect these various aspects of the subject from all levels of undergraduate teaching into the upper bracket of research.

BONDING AND STRUCTURE
Structural Principles in Inorganic and Organic Chemistry

N. W. ALCOCK B.A., Ph.D.
Reader in Chemistry
University of Warwick, Coventry

ELLIS HORWOOD
NEW YORK LONDON TORONTO SYDNEY TOKYO SINGAPORE

First published in 1990 by
ELLIS HORWOOD LIMITED
Market Cross House, Cooper Street,
Chichester, West Sussex, PO19 1EB, England

A division of
Simon & Schuster International Group
A Paramount Communications Company

© Ellis Horwood Limited, 1990

Printed and bound in Great Britain
by Hartnolls, Bodmin, Cornwall

British Library Cataloguing in Publication Data

Alcock, N. W. (Nathaniel Warren) 1939–
Bonding and structure: structural principles in inorganic and organic chemistry.
1. Chemical bonding
I. Title
541.224
ISBN 0–13–465253–3 (Library Edn.)
ISBN 0–13–465246–0 (Student Edn.)

Library of Congress Cataloging-in-Publication Data

Alcock, N. W. (Nathaniel Warren)
Bonding and structure: structural principles in inorganic and organic chemistry /
N. W. Alcock.
p. cm. — (Ellis Horwood series in inorganic chemistry)
Includes bibliographical references and index.
ISBN 0–13–465253–3 (Library Edn.)
ISBN 0–13–465246–0 (Student Edn.)
1. Chemical bonds. 2. Molecular structure. I. Title. II. Series.
QD461.A385   1990
541.2'24–dc20   90–43931
                                                                      CIP

Table of contents

To all students of chemistry

Preface

At the atomic level, electrostatic forces alone are responsible for the bonds in chemical elements and compounds. However, these forces are manifested in a great variety of forms, from the delocalized electrons in sodium to the van der Waals contacts in a noble gas. Conventionally, chemistry texts have concentrated on covalent and ionic compounds. It is my strong belief that bonding should be considered as a unity. The dependence of bonding, of structure, and of properties on atom type can then be examined for every sort of material. This book is the result. To this aim, it includes some material more often found in physics than in chemistry textbooks, but also accompanies bonding in inorganic compounds with the discussion of what are conventionally organic compounds. I hope the opportunity to see the whole picture will make it easier for students to fit their specific studies into the wider view. The recent developments, both within the branches of chemistry and between chemistry and physics, of interdisciplinary studies of synthesis, properties and technological applications of novel materials should make this approach particularly appropriate for students of chemistry in the 1990s.

The second inspiration for writing has come from the problems that students inevitably encounter in studying structure and bonding. Experience as a lecturer and tutor has shown the importance of presenting the principles of bonding in a logical sequence, and has helped me find new answers to a few of the problems. As will become clear, I am building on the work and inspiration of other chemists, from Linus Pauling's trail-blazing *The nature of the chemical bond* onwards. I hope these developments of earlier studies will prove useful to students, and perhaps occasionally to their lecturers. I believe it is unnecessary to overwhelm students with mathematical formulae. I have therefore adopted a non-mathematical approach, as far as possible. Most of the equations relating aspects of properties and bonding are given without derivation, though appendices to some chapters provide mathematical details, for readers interested in exploring further.

Many debts have been incurred during the writing of this book. I am particularly appreciative of the Research School of Chemistry, Australian National University, where an appointment as a Visiting Fellow gave the opportunity to escape daily responsibilities while assembling the book's text in their excellent library. Many colleagues at Warwick have shared their expertise with me, and I hope will not be alarmed to find their idle responses included here; the errors are, of course, my own. The students, past, present, and future, to whom the book is dedicated, deserve special mention. May they never cease by their questions to force their teachers to remain students themselves.

N. W. Alcock

A note to the reader
Certain important topics may be familiar to some readers. These are described in self-contained units, placed in boxes when the topic is first encountered; the boxes are listed at the end of the Table of Contents. A few other points, including mathematical derivations, which are useful but not essential for the main sequence of ideas, are placed in chapter appendices. The main appendix (p. 315) tabulates values of radii appropriate for the different bonding types.

Part I
The diversity of bonding

1

Themes and variations

1.1 INTRODUCTION

The chemical elements and compounds are amazing in their variety. From helium a gas at 4.2 K, to W_2C a solid at 3400 K, from MgO an almost perfect insulator, to $YBa_2Cu_3O_7$ a superconductor at 90 K, from the softness of sodium, to the strength of diamond or zirconia, their diversity of properties reflects their equally diverse bonding and structure. The aim of this book is to explore chemical bonding in all its forms, to follow the path from each type of bonding to the structures with which it is associated, and to outline the relationship with its characteristic properties.

Simply because inorganic compounds show a far wider range of bonding than found in most organic compounds, this is principally a book on inorganic chemistry. 'Organic' compounds, of course, have equally significant relationships between their bonding and their structures, and so they find their places in the appropriate chapters. This juxtaposition illustrates the underlying unity of chemistry, and the opportunity to compare 'inorganic' and 'organic' bonding reveals links that are often overlooked. Similarly, bonding in the solid state dominates several chapters, although molecular compounds are not neglected. This emphasis arises naturally, because many types of bonding exist only in solids, and it gives an opportunity to explore the fascinating recent discoveries of inorganic chemistry, in such fields as superconductivity, ferromagnetism, and intercalated metal compounds.

Part I provides a guide for classifying the diversity of bonding types by relating them to one property of the bonded atoms, their *electronegativity* (the ability of an atom in a compound to attract electrons). The *bond triangle* (section 1.2.1) shows the separation by electronegativity of the three main types of bond. It also reveals a middle region that is often ignored, containing compounds with intermediate bond character. Electronegativity itself is a chemical concept of fundamental importance in the systematic correlation of bond types. Unfortunately, it is also one of the least satisfactory atomic parameters, because of the difficulty of producing a quantitative scale for it. It is considered in detail in section 1.3.

It is always important to assess the validity of inferences about bonding, in relation to the available sources of evidence. Chapter 2 therefore outlines how the structure, bonding, and stability of compounds can be studied. It is not a detailed guide to the available techniques, but rather provides the framework for such a guide. Both in this chapter and later, the emphasis is on the determination and interpretation of molecular structure and dimensions. Although it is sometimes difficult to understand the significance of such evidence for bonding, comparison with carefully chosen model compounds generally gives trustworthy information on bond type and strength. Standard values for the radii of atoms forming metallic, ionic, or covalent bonds are very useful for comparison with observed bond lengths. Their derivations from the observed structures are discussed in Chapters 3–5, and the values are collected in the appendix (p. 315).

1.2   A GUIDE THROUGH THE MAZE

If we had a perfect understanding of chemical behaviour, we would be able to predict the chemical and physical properties of a given compound from the fundamental properties of its component atoms. In general this ideal is far beyond us, though it is possible to calculate the overall stability of simple covalent compounds from first principles. For more complex systems, the most that can be achieved is to correlate chemical behaviour with bond type. In turn, the bond type can be related to the properties of the atoms forming the bond. Because the bonding involves the interaction of the valence electrons with the bonded atoms, the strength of the interaction between the electrons and the atoms plays a dominant part in determining the nature of the bonding. It is therefore most useful to take this atomic property, the *electronegativity* (section 1.3), as a guide to bonding. The relationship is most clearly demonstrated for binary compounds A_mB_n, and can be explored in the bond triangle.

1.2.1   The bond triangle

If we allow both A and B to vary in the compounds A_mB_n, from the least electronegative (or most electropositive) element, which has the least ability to attract electrons, to the most electronegative element, we cover all the major types of bonding. The relationship between electron attraction and bonding can be displayed in a square, with A varying along one side and B along the other, both ranging from low to high electronegativity (Fig. 1.1a). By always taking B to have the higher electronegativity, we need only consider half the square (Fig. 1.1b). The result is the bond triangle, which gives the framework to this book.

The ideal types of bonding occupy the corners of the triangle (Fig. 1.2). With A and B both electropositive, neither attracts electrons strongly. Electrons are therefore lost from the individual atoms and move freely within the solid, held by the positively charged atoms. This mutual attraction gives *metallic bonding* (Chapter 3). If atom A is electropositive and B very electronegative, electrons easily lost by A are strongly attracted to B. The resulting ions, A^+ and B^-, are held together by electrostatic attraction in *ionic bonding* (Chapter 4). Finally, if both A and B are both electronegative, neither can lose electrons. *Covalent bonding* is found, in which the valence electrons are held by both atoms (Chapter 5).

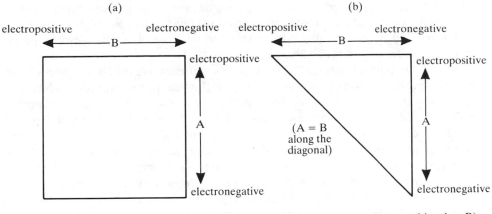

Fig. 1.1 — (a) Variation of electronegativity in a binary compound A_mB_n; (b) variation of electronegativity in A_mB_n when B has the higher electronegativity.

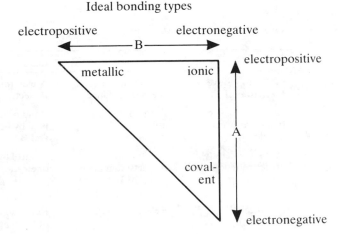

Fig. 1.2 — Ideal bonding types.

Part II investigates these three major bond types at a simple level. Each chapter examines ideal bonds of the appropriate type, and determines the consequences for structure, for properties, and for energy. Each bond type imposes distinctive geometries on its compounds. Thus, in metallic bonding, the greatest stability is achieved when the material has maximum density, i.e. the free electrons are most concentrated. Ionic compounds also take up dense structures, with as many short A^+

to B⁻ distances as possible, but the repulsions between ions of like charge must be
minimized. In contrast, covalent compounds do not generally have high densities, as
their geometry is controlled by the number of pairs of electrons forming bonds
between one atom and its neighbours.

Another important distinction between these different types of bonding concerns
their binding energies. For a stable compound, the heat of formation can always be
measured experimentally. However, it is more valuable for interpreting chemical
reactivity and properties, to understand how the bond energy is related to the
properties of the component atoms. This is easy for ionic compounds. As the ionic
forces result from simple electrostatic interactions, their energies can easily be
calculated from the sizes of the ions. The relationship between size and energy can
then be used to give accurate predictions of the behaviour of ionic compounds. For
covalent compounds, a qualitative measure of bond strength can also be obtained
easily. This is the *bond order*, given by the number of bonding pairs of electrons. To
calculate bond energies quantitatively requires sophisticated quantum-mechanical
calculations. It is possible only for fairly simple molecules, and even then requires
very lengthy computations. For metallic bonding, empirical correlations between
atom size and energy are possible, but even sophisticated computations cannot yet
reproduce observed energies very successfully.

Table 1.1 summarizes the key features of these ideal bond types. It is worth

Table 1.1 — Summary of ideal bond types

Type	Metallic	Ionic	Covalent
Formed by	A electropositive B electropositive	A electropositive B electronegative	A electronegative B electronegative
Nature of bonding	Attraction between free electrons and metal ions	Electrostatic attraction of A^+ and B^-	Bonding electrons held by both A and B
Geometry	Maximum density of electrons and metal atoms	Maximum A–B contacts; minimum A–A, B–B contacts	Control by number of electrons
Energy : Qualitative comparisons	Empirical correlation	Relate to ion size	Relate to number of bonding electrons
Quantitative calculation	Difficult and not satisfactory	Simple electro- static energy calculations	Quantum-mechanical calculations for simple molecules

noting that metallic and ionic bonding require 3-dimensional networks of interactions. They are therefore found in solids and to some extent liquids, but not gases.
Covalent bonding also exists in some extended solids, but is the only type found in
small molecules; these may be gases, liquids, or solids at normal temperatures, but in
their condensed phases the covalent bonds are accompanied by non-covalent van der
Waals forces (Chapter 7).

The elements (A or B alone) lie along the diagonal of the bond triangle. Thus, they can show metallic or covalent bonding or intermediate types, but cannot be ionic.

1.2.2 Intermediate bonding

Not all compounds lie at the corners of the bond triangle and have one of the ideal bond types. It is more accurate, though something of an exaggeration, to suggest that *no* 'real' compounds have ideal bonds. At the same time as the bonding becomes 'real', so the properties become more interesting, often with important technological applications. Two chapters in Part III extend the descriptions of metallic and ionic compounds beyond those of Part II (Chapters 6,10). The problems found with 'real' covalent compounds (Chapter 9) more often arise from the complicated bond interactions, rather than from departures from covalent character. In addition, a variety of forces are found *between* covalent molecules. Chapter 7 therefore examines both van der Waals forces and *dative* bonding (including the varieties known as hydrogen bonding and secondary bonding). In Chapter 8, the important interaction between a metal ion and a covalent molecule or ion is considered. It is worth noting that as a consequence of its formation by pre-existing molecules, dative bonding is the only type not found in binary compounds.

The final chapter (11) explores the centre of the bond triangle: compounds whose bonding shares the characteristics of two or even all three main types; this area includes technically important classes of compounds, such as semiconductors and silicates, but also covers strange and unusual materials, like the Zintl phases (e.g. $CaSi_2$) or the compounds of graphite (C_8Br). Fig. 1.3 shows the places of these intermediate and modified bond types in the bond triangle.

1.3 ELECTRONEGATIVITY

Electronegativity was originally defined by Linus Pauling as 'the power of an atom in a molecule to attract electrons to itself'. Inevitably, this definition leads to ambiguity, because an essentially *atomic* property is meaningful only if the atom concerned is part of a molecule. It is unclear whether the electroncgativity should be constant, or may be influenced by the other atoms in the molecule. Furthermore, the definition gives no guidance about ways of measuring electronegativity.

Pauling himself developed a scale of electronegativity based on energy data. For a bond A–B, this scale uses the difference between the A–B bond energy and the mean of the A–A and B–B bond energies. The difference is attributed to the ionic contribution to the bonding ($A^+ B^-$ or $B^+ A^-$); the extent of this contribution depends on the electronegativity difference between A and B. Choice of suitable scale factors gave electronegativity values in the range 1.0 (Li) to 4.0 (F). Since then, the relationships have been studied between these values and many chemical and physical parameters, for example nuclear magnetic resonance chemical shifts for protons in H–C–X bonds, and the acid strengths of H–X or H–C–X groups. Many of these parameters show good correlations with electronegativity. These and other quantities have therefore been used in reverse, to construct scales of electronegativity.

Two scales seem to be more soundly based and more comprehensive than the

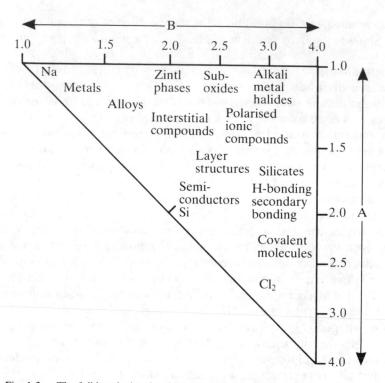

Fig. 1.3 — The full bond triangle with intermediate as well as ideal bonding types.

alternatives. The first is that of Mulliken. He argued that the competition in a compound AB between the ionic forms A^+B^- and A^-B^+ is controlled by the relative electronegativities of A and B. The energies of the two ionization processes are:

(ionization potential of A) + (electron affinity of B)

and the reverse of this. The electronegativity of each atom can therefore be given by (ionization potential) + (electron affinity) (taking both as positive quantities). As well as the simple Mulliken electronegativities for atoms in their ground states, values can also be obtained for atoms in different oxidation states and hybridizations.

The second scale is that of Allred and Rochow (see *References*). They calculated the force exerted by the nuclear charge, at a distance corresponding to the covalent radius of the atom(r). Thus,

$$\text{Force} = z_{\text{eff}}/r^2 .$$

The effective nuclear charge, z_{eff}, makes allowance for screening by the outer electrons. In principle these values also depend on oxidation state and hybridization (Chapter 5) (both of which affect covalent radius), though figures have been calculated only from single bond radii for atoms in their most common oxidation state.

It is remarkable that both the Mulliken and Allred & Rochow scales produce good straight lines when plotted against Pauling's original electronegativities (after adjusting the overall ranges to correspond). Even though the theoretical justifications for the scales are very different, these correlations do suggest that they are measuring a real atomic property.

The selection of one specific scale for this book is somewhat arbitrary, but the Allred & Rochow has been chosen, because it is probably the most widely used. With the appropriate scaling factors to convert it to the Pauling scale, it becomes

$$\chi(\text{electronegativity}) = 0.359 z_{\text{eff}}/r^2 + 0.744.$$

The resulting values are tabulated in Fig. 1.4. It is worth noting that they are far from evenly distributed, with only the most electronegative elements having electronegativities exceeding 2.0, and the majority of metals clustering in the 1.3 to 1.6 range; for this reason, a non-linear electronegativity scale is shown in Fig. 1.3. The principal trends of electronegativity within the periodic table are an increase from

1 IA	2 IIA	3 IIIA	4 IVA	5 VA	6 VIA	7 VIIA	8 ←VIII	9 VIII	10 VIII→	11 IB	12 IIB	13 IIIB	14 IVB	15 VB	16 VIB	17 VIIB
H 2.2																
Li 1.0	Be 1.5											B 2.0	C 2.5	N 3.1	O 3.5	F 4.1
Na 1.0	Mg 1.2											Al 1.5	Si 1.7	P 2.1	S 2.4	Cl 2.8
K 0.9	Ca 1.0	Sc 1.2	Ti 1.3	V 1.5	Cr 1.6	Mn 1.6	Fe 1.6	Co 1.7	Ni 1.8	Cu 1.8	Zn 1.7	Ga 1.8	Ge 2.0	As 2.2	Se 2.5	Br 2.7
Rb 0.9	Sr 1.0	Y 1.1	Zr 1.2	Nb 1.2	Mo 1.3	Tc 1.4	Ru 1.4	Rh 1.4	Pd 1.4	Ag 1.4	Cd 1.5	In 1.5	Sn 1.7	Sb 1.8	Te 2.0	I 2.2
Cs 0.9	Ba 1.0	La-Lu ~1.1	Hf 1.2	Ta 1.3	W 1.4	Re 1.5	Os 1.5	Ir 1.6	Pt 1.4	Au 1.4	Hg 1.4	Tl 1.4	Pb 1.6	Bi 1.7	Po 1.8	At 2.0
Fr 0.9	Ra 1.0															

Lanthanides ~ 1.1

Fig. 1.4 — Electronegativity values for the element. Note: both new (1–18) and old (I–VIII) group numbers are given. Source: Allred & Rochow (1958) (rounded to one decimal place).

left to right in each period, and a decrease down each group. One anomaly should be noted. The presence of the transition metals produces a decrease in size for the elements following them, and therefore increases their electronegativity. As a result, the electronegativity of Ge on this scale exceeds that of Si. The general chemical behaviour of silicon is more realistically represented by an electronegativity of about 2.0 to 2.1; in particular, this differentiates it from the transition metals, and equates it more closely with boron.†

† This anomaly has been closely examined by Allred, A. L. & Rochow, E. G. (1958) 'Electronegativities of carbon, silicon, germanium, tin and lead'. They show that the alternation between Si and Ge is consistent with the chemistry of these elements; the *relative* electronegativity values down this group are therefore likely to be correct. However, they did not investigate the relationship between these values and those of other groups, or the wider implications of their values. In particular, it is not clear if the radii used in calculating the electronegativities for the transition metals are consistent with those for the non-metals.

Many sophisticated calculations of electronegativity values have been carried out, including the assignment of values to groups, as well as to atoms in different valence states. In this book, their most important purpose is to give an impression of the chemical character of each element; the values in Fig. 1.4 should achieve this.

FURTHER READING

Allred, A. L. & Rochow, E. G. (1958) 'A scale of electronegativity based on electrostatic force', *J. Inorg. Nucl. Chem.* **5** 264.

Sen, K. D. & Jørgensen, C. K. (eds) (1987) 'Electronegativity', *Structure & bonding*, **66**. This gives a full analysis of modern studies of electronegativity scales and applications.

2

The evidence

This short chapter does not attempt to describe in detail the numerous techniques applied in the study of structure and bonding. Several books do just that, both at elementary and advanced level (see *Further reading*). Instead, its aim is to point out what *can* and what *cannot* be discovered about structure and bonding, and identify the methods needed. In a way, it therefore acts as an index to a more detailed book. The emphasis is on methods giving quantitative information on structures, and especially on X-ray and neutron diffraction. This needs an explanation, though not an apology. Much of the structural evidence used in later chapters relates to solids and can be obtained only by these techniques. For covalent compounds, information about the overall structure is also valuable. However, even with these compounds, knowledge of the precise dimensions adds greatly to our understanding of their bonding.

In studying the structure and bonding of any material, we require two items of information in relation both to its atoms and its electrons. We need to know their POSITIONS and their ENERGIES. Thus four aspects are involved:

(a) the positions of the atoms in the overall structure,
(b) the bond energy of the molecule as a whole and of individual bonds in it,
(c) the distribution of the electrons over the molecule and the identity of the orbitals they occupy,
(d) the energies of the individual electrons (in their orbitals).

2.1 MOLECULAR STRUCTURE AND ATOMIC POSITIONS

The structure of a molecule can be determined at three levels of sophistication.

(i) At the simplest level, the *connectivity* of the molecule can be established, showing which atoms are linked to which. For example in Fig. 2.1a, *A* and *B* can be distinguished from their connectivity. Connectivity can usually be found by spectroscopic methods, principally nuclear magnetic resonance (n.m.r.) and infrared (i.r.) spectroscopy, helped by mass spectrometry. For a rather rigid molecule, the relative

(a)

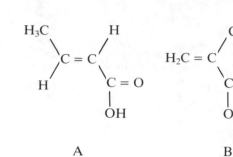

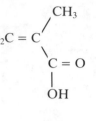

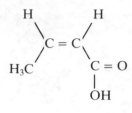

A   B   C

(b)

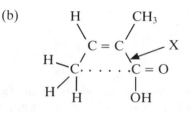

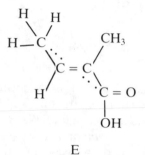

D   E

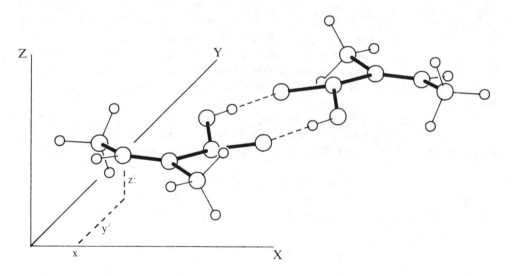

Fig. 2.1 — (a) Molecular connectivity information (b) Molecular distance information (c)
Crystal coordinate information.

arrangement of different parts of it may also emerge from these techniques; the *cis-*
and *trans-*isomers, A and C (Fig. 2.1a) could be distinguished.

The connectivity is an essential aspect of molecular structure, but for detailed
examination of bonding, it needs to be supplemented by quantitative information.

(ii) Some techniques establish *interatomic distances* (Fig. 2.1b) (Box 1). These usually measure distances between neighbouring atoms quite precisely. For example in D it would be found that the C–C bond (X) was unusually short because of the influence of the neighbouring C=C and C=O bonds. The distances between atoms that are slightly further apart can also be determined, but not those at longer range. Thus, the dotted C...C distances distinguish the isomers C (angelic acid) and D (regrettably not *diabolic* but tiglic acid). The preferred shape of the molecule would remain uncertain; one could not discover if all four carbon atoms tended to lie in one plane.

(iii) The fullest information comes from the determination of the 3-dimensional coordinates (x, y, z) for every atom in the molecule (Fig. 2.1c). These coordinates can obviously be used to calculate the distances between all bonded atoms, but they will also give torsion angles about bonds and longer intramolecular and intermolecular contacts such as the O–H ... O hydrogen bonds (section 7.4) in D. Unlike the evidence of type (ii), the precision of these calculations does not decrease as the distances increase.

One crucial difference exists between type (ii) and type (iii) techniques. The latter can be applied only to crystalline materials, in which all the molecules have specific constant orientations. Information of type (ii) can be obtained from gases, liquids, and amorphous solids in which the molecular orientations are random. It is so much more informative to determine coordinates rather than distances, that crystal structural methods (iii) are used if at all possible. Other techniques are employed only if crystals cannot be obtained, or if (say) the comparison of gas phase and solid state structures is required.

2.1.1 Crystals

The importance of crystals in structural studies has led to many of the terms used to describe crystals being applied also to the structures. The starting point for these descriptions is the crystal itself. Although the most obvious characteristics of most crystals are their gleaming faces, these are not their most important features. Some crystals lack any well-defined faces, and occasionally objects with excellent faces prove to be amorphous (as crystallographers know to their regret). The essential feature of a crystal is its *internal regularity*, not its external form.

Box 1 Units used for bond lengths

This book uses picometres as the unit for distances between atoms (1 $pm = 10^{-12}$ m). Distances are usually determined to a precision of at least 1 pm, and occasionally to 0.1 pm, and the error can be assumed not to be significant for the distance as stated.

An older unit, the Ångstrom, is still very common, especially in crystallographic literature $1\text{Å} = 10^{-8}$ cm or 10^{-10} m. The nanometre is also occasionally used (1000 pm = 1 nm). As an example, a typical C–C distance is 153.0 pm (= 1.530 Å = 0.1530 nm).

A crystal is made up by the repetition in three dimensions of a block, the *unit cell* (Fig. 2.2) (Box 2). Each unit cell contains one or more molecules (generally more than one), and all unit cells in the crystal are identical. The shape of the unit cell is characteristic of the material involved, but the unit cells can be grouped into families (known as *crystal systems*) according to how symmetrical the crystals are. The names of these families are frequently used in describing structures, e.g. 'hexagonal' close-packed metals or 'monoclinic' sulphur, so it is important to have some familiarity with them.

If the crystals contain no symmetry at all, then the unit cell is an oblique block (a *parallelepiped*), described by the six parameters shown in Box 2. Crystals of this system are known as *triclinic*. As the symmetry of the crystal increases, so the number of parameters needed to describe the unit cell decreases. Thus, if the unit cell contains one 2-fold axis of symmetry, two of the angles (usually taken as α and γ) are forced to be 90°, leaving four variable parameters, the unit cell 'constants'. This unit

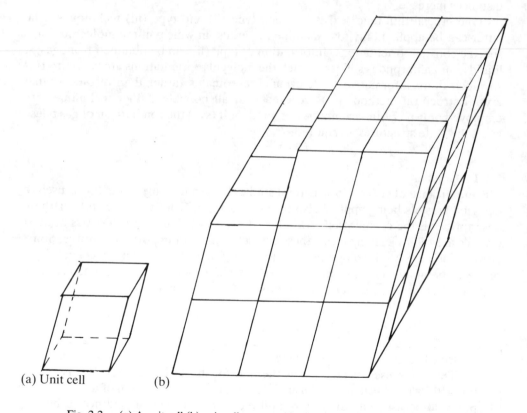

(a) Unit cell   (b)

Fig. 2.2 — (a) A unit cell (b) unit cells repeated in three dimensions to give a crystal.

cell belongs to the *monoclinic* system. Table 2.1 summarizes the seven crystal systems, their symmetry, and the number of variable unit cell parameters for each.

A further term often found in descriptions of crystals is 'lattice' or 'crystal lattice',

Box 2 Unit cell parameters and fractional coordinates

Six parameters are needed to describe a unit cell (Fig. B2(a)): a, b, c, the lengths of the sides of the block, and α, β, γ, the angles between the sides (with α being the angle between b and c, β between a and c, and γ between a and b).

To specify the positions of atoms in a crystal, special sets of coordinates are generally used, known as *fractional atomic coordinates* x, y, z (Fig. B2(b)). Each atom is located by moving parallel to each axis of the unit cell, as a fraction of the length of that axis. Thus an atom at 0.6, 0.5, 0.4 is reached from the origin by moving 0.6 a parallel to a, 0.5 b parallel to b, 0.4 c parallel to c. The convenience of these coordinates is that they correspond precisely to the repetitions of the unit cell in the crystal. Thus, the atom at 1.6, 0.5, 0.4 is at exactly the same position as the last one, but in the next unit cell in the a direction. As each unit cell is identical, adding an integer to any of the coordinates just moves to the corresponding atom in another cell.

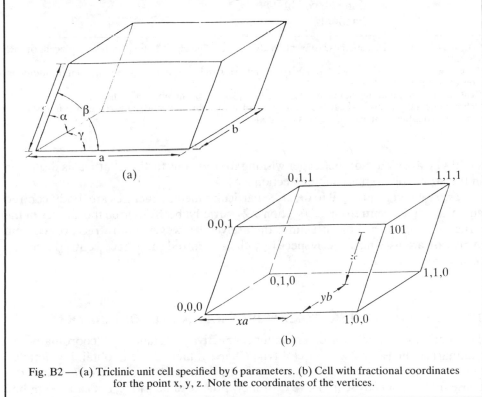

Fig. B2 — (a) Triclinic unit cell specified by 6 parameters. (b) Cell with fractional coordinates for the point x, y, z. Note the coordinates of the vertices.

in phrases such as 'The crystal lattice of NaCl contains ...'. Unfortunately, lattice has a precise and rather subtle meaning in relation to crystals, as 'a mathematical array of points'. Its use as a variation on 'unit cell' or 'crystal structure' leads to confusion, and

Table 2.1 — Crystal systems[a]

System	Symmetry axes[b]	Variable parameters	Constraints on parameters
Triclinic	none	$6\ (a,\ b,\ c,$ $\alpha,\ \beta,\ \gamma)$	none
Monoclinic	2-fold axis along b	$4\ (a,\ b,\ c,\ \beta)$	$\alpha=\gamma=90°$
Orthorhombic	three 2-fold axes	$3\ (a,\ b,\ c)$	$\alpha=\beta=\gamma=90°$
Tetragonal	4-fold axis along c	$2\ (a,\ c)$	$a=b,\ \alpha=\beta=\gamma=90°$
Hexagonal[c]	3- or 6-fold axis along c	$2\ (a,\ c)$	$a=b,\ \alpha=\beta=90°$ $\gamma=120°$
Rhombohedral[d]	3-fold axis along cell body diagonal	$2\ (a,\ \alpha)$	$a=b=c,$ $\alpha=\beta=\gamma$
Cubic	four 3-fold axes along all cell body diagonals	$1\ (a)$	$a=b=c,$ $\alpha=\beta=\gamma=90°$

[a]The axis directions used are those conventionally adopted. See Fig. 2.10 for pictures of unit cells of each system.
[b]These are the *essential* symmetry axes. Crystal systems with high symmetry generally contain additional axes of symmetry.
[c]Note particularly that the 'hexagonal' unit cell is in plan a 120° rhombus, *not* a hexagon.
[d]Rhombohedral crystals can be redefined to have hexagonal unit cells, and so the rhombohedral system is sometimes omitted from lists of the crystal systems.

should be avoided. For the reader wishing to study this further, 'lattice' is discussed in the Appendix to this chapter (section 2.5).

Two related terms used to describe particular metal structures are 'body-centred cubic' and 'face-centred cubic' (section 3.2). Strictly, both describe the lattices of the structures rather than the structures themselves. However, both structures contain arrays of atoms which are respectively 'body-centred' and 'face-centred', so the terms are not incorrect.

2.2   OBTAINING INTERATOMIC DISTANCES AND COORDINATES

The methods available for determining connectivity, distances, or coordinates are summarized in Table 2.2. Those of type (i) are of limited use in obtaining detailed bonding evidence. Among the type (ii) methods, the first two (microwave and gas-phase infrared spectroscopy) give extremely precise dimensions, but can be applied only to simple volatile molecules. Apart from EXAFS (section 2.2.4), the remaining techniques fall within the broad category of *diffraction methods*. Their essential character is illustrated in Fig. 2.3. When a beam of X-rays (or neutrons or electrons) strikes two atoms, the scattered X-ray beams in a particular direction will be out of phase with each other to an extent depending on: the wavelength of the X-rays, the distance between the atoms, and their orientation. This phase difference causes the

Table 2.2 — Structural techniques

Type (i): Information on connectivity but not distances

Technique	Information	Restrictions
Mass spectrometry	Mass and fragmentation pattern	Connectivity obtained indirectly
Nmr spectroscopy	Environment of magnetic nuclei	Some nuclei inaccessible
Infrared spectroscopy: solid	Characteristic groups	Connectivity obtained indirectly

Type (ii): Information on distances but not coordinates

Technique	Interaction with	Restrictions
Microwave spectroscopy	Rotational energy levels. Very precise dimensions	Volatile compounds with few atoms
Infrared spectroscopy: gas-phase	Rotational energy levels	Similar to microwave but more restricted
Gas phase electron diffraction	Charge of electrons	Moderate sized molecules (maximum about 20 atoms); must have some volatility
X-ray and neutron diffraction of liquids	Electrons	Little used. Has provided some information on pure liquids but difficult to apply to dissolved species.
Extended X-ray absorption fine structure: EXAFS	Electrons	Provides distances from one type of atom (heavier than C) Material need not be crystalline. Normally needs synchrotron X-ray source.

Type (iii): Information on coordinates

Technique	Interaction with	Restrictions
X-ray diffraction: single crystal	Electrons, so insensitive to atoms with few electrons (e.g. H)	Material must be individual crystals, normally at least 0.1 mm in one dimension Inconvenient for material melting below room temperature.
X-ray diffraction: powder	Same	Material can be micro-crystalline; easier for low-melting solids Difficult to determine coordinates for compounds of more than moderate complexity.
Neutron diffraction: single crystal or powder	(a) nucleus - not related to number of electrons (b) unpaired electrons (magnetic scattering)	Requires high intensity source of neutrons (e.g. nuclear reactor) Other restrictions as for X-ray methods
Electron diffraction (crystalline solids)	Charge of electrons	Electrons strongly absorbed, so very thin samples needed, but uses very small area (down to 1 nm square). Materials must be involatile Determination of structure difficult if material of more than moderate complexity.
Low-energy electron diffraction (surfaces of crystals) LEED	Same	Interaction only with surface atomic layers. Also gives information on layers of adsorbed molecules. Determination of structures and positions of adsorbed molecules difficult

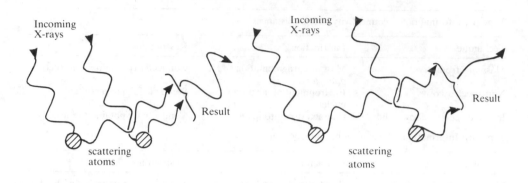

Fig. 2.3 — Interaction between beams of X-rays. Waves scattered from two atoms undergo interference. Depending on the spacing between the atoms and the direction of the rays, the waves may combine to produce a strong (a) or a weak (b) diffracted beam.

rays partly to cancel each other out, reducing the intensity of the scattered beam. A different distance or orientation (Fig. 2.3b) produces a different phase difference and intensity. The intensity of the scattered or 'diffracted' beam is therefore controlled by the arrangement of atoms in the molecule diffracting the X-rays. For structure determination, the intensities of the scattered X-ray beams are measured and the atomic arrangement worked out from the intensities.

2.2.1   X-ray single crystal diffraction

X-ray diffraction from single crystals is by far the most important of all these structural techniques, with about 10 000 structures reported every year. The X-rays are scattered by the electrons in the atoms, and a portion of a typical map of the electron density, produced by X-ray diffraction is shown in Fig. 2.4a; from this, the atomic positions can readily be determined.†

The interaction of the X-rays with the electrons leads to one important limitation of this technique. In a molecule containing 'heavy' atoms (i.e. those of high atomic number with many electrons), these will influence the scattering more than the 'light' atoms. As a result, calculation of atomic positions from the scattered intensities gives more precise coordinates for the heavy atoms than for the light ones. As an example, in a recent determination of the structure of $(CH_3)_2S$-Au-Cl, the error in the gold atom coordinates was $\pm$ 0.03 pm, that in sulphur and chlorine 0.2 pm, while for carbon the error was 0.9 pm. (Jones & Lautner 1989).

Hydrogen atoms are particularly affected by this problem, so that even in a compound of C and H only, with typical errors in C-atom positions of 0.5 pm, the H-atom errors might be 5 pm. In compounds also including heavier atoms, even the location of H-atoms may be difficult. Their positions are therefore often assumed.

---

† In the early days of X-ray diffraction, such maps were studied with great care, but now the values of electron density at an array of points in the crystal are generally stored in computer memory and searched for maxima corresponding to atomic positions. These positions are then manipulated directly.

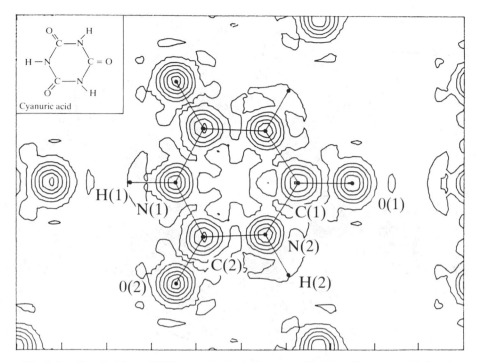

Fig. 2.4 — Density Maps: (a) Electron density in the molecular plane of cyanuric acid (inset) from X-ray diffraction. Note the low electron density of the hydrogen atoms. Source: calculated from data in Verschoor, G. C. & Keulen, E. (1971) *Acta Cryst.* **B27** 134.

For example, in a -CH$_2$- group, the C–H distance might be taken as 96 pm, and the H–C–H angle as 109.5°. In examining distances involving light atoms, it is therefore important to take note of the precision of the values (see, for example, the discussion of hydrogen bonding in section 7.4)

2.2.2 Neutron diffraction

The major limitation of X-rays in locating light atoms is overcome in the technique of *neutron diffraction*. Neutrons are scattered by interaction with the atomic nuclei, rather than the electrons. The extent of this scattering is controlled by the nuclear structure, but is unrelated to the number of protons (the atomic number). Table 2.3 lists a selection of neutron scattering 'lengths' and the corresponding X-ray values. The range of neutron scattering lengths is small, so that almost any type of atom can be accurately located by neutron diffraction. In particular, H-atoms can be found even in the presence of heavy atoms. Atoms with almost the same numbers of electrons such as Fe and Co, which are virtually indistinguishable by X-ray techniques, can also easily be identified when using neutrons.

One feature unique to neutrons is that they have a magnetic moment. They are

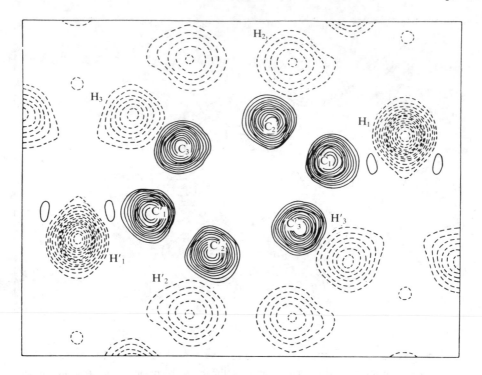

Fig. 2.4 — Density Maps: (b) Neutron scattering density of benzene, with negative contours
dashed. Source: Glusker & Trueblood (1985) p. 178.

Table 2.3 — Comparison of neutron and X-ray scattering

Neutron scattering is expressed in units of length, while the units of X-ray scattering
are electrons, giving values relative to the scattering from an atom containing one
electron.

Atom	Neutron scattering[a]	X-ray scattering
	$(10^{-12}$ cm$)$	(els)
H	-0.37	1
D	0.67	1
C	0.66	6
V	-0.04	23
Fe	0.95	26
Co	1.03	27
Pt	0.95	78
U	0.85	92

[a]The negative values for some neutron scattering lengths merely indicate that scattering from these atoms
involve a reversal of the phase of the scattered beam. The near-zero value of vanadium's scattering length
means that vanadium atoms are almost impossible to locate with neutron diffraction.

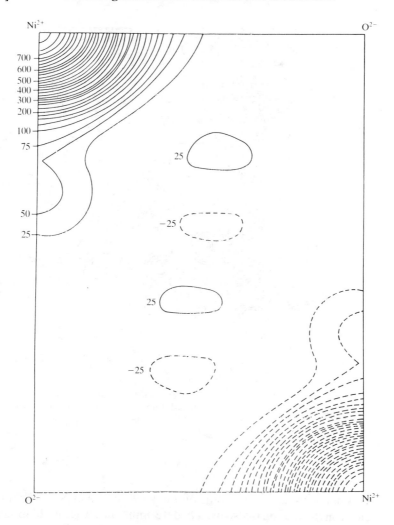

Fig. 2.4 — Density Maps: (c) density of a magnetic electron in NiO from neutron scattering.
Source: reproduced with permission from Alperin, H. (1962) *J. Phys. Soc. Japan.* **17** B-III, 12.

therefore scattered by the magnetic moments of unpaired electrons. Neutron
diffraction is the only technique useful for identifying the complicated arrangements
these electrons often take up in magnetic materials (section 10.8). In favourable
cases, it can even reveal the distribution of a single unpaired electron in a structure
(Fig. 2.4c).

2.2.3 Electron diffraction

The two electron diffraction techniques of type (iii) listed in Table 2.2 are similar in
principle to X-ray and neutron diffraction, but are specialized in their application,

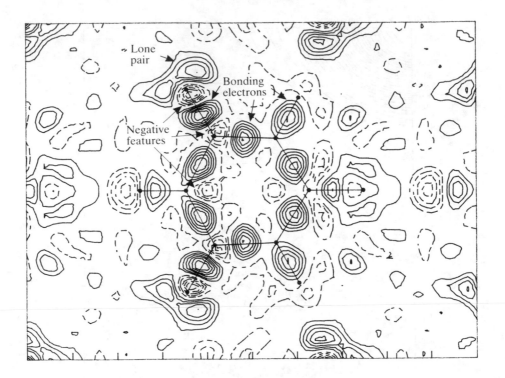

Fig. 2.4 — Density Maps: (d) 'deformation' electron density, showing anomalies caused by bonding electrons in cyanuric acid. Source: calculated from data in Verschoor, G. C. & Keulen, E. (1971) *Acta Cryst.* **B27** 134.

either to very small crystals, or to atoms on the surface of a crystal. As a type (ii) technique, electron diffraction is also used to determine the structure of molecules in the gas phase. The random orientation of the molecules means that the diffraction effects illustrated in Fig. 2.3 are observed simultaneously for molecules in all orientations. As a result, the diffraction intensities can only be converted into a plot showing the distances between scattering atoms, a *radial distribution function* (RDF). The plot for acetone is shown in Fig. 2.5a. The first peak corresponds to the six C–H distances, but this overlaps with the C=O distance, and partly with C–C. Beyond these come the distances between atoms not directly bonded. As Fig. 2.5a shows, even for such a simple molecule the RDF is complicated and difficult to interpret.

2.2.4   EXAFS: extended X-ray absorption fine structure

EXAFS (extended X-ray absorption fine structure) is based on the fluctuations in the absorption of X-rays by atoms of one particular type, caused by the presence of neighbouring atoms. This technique also generates an RDF, but gives rather clearer structural information because it includes only the distances from the central atom.

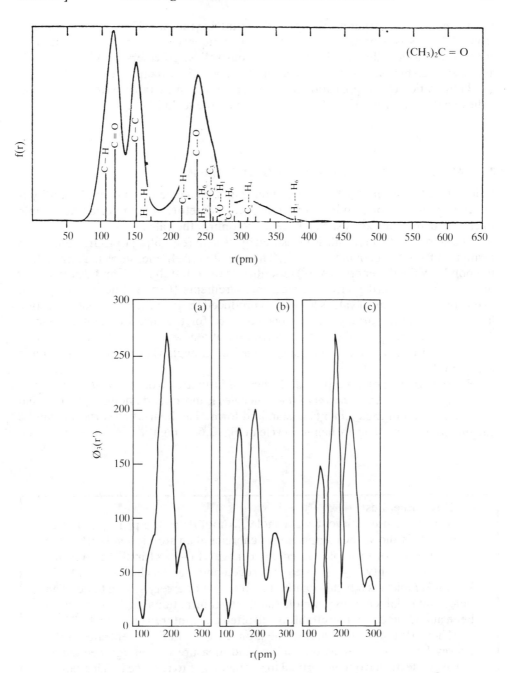

Fig. 2.5 — (a) Radial distribution function for acetone, marked with all distances (b) EXAFS radial distribution functions (Fe) for two Fe–Mo–S model clusters and the Fe–Mo cofactor of nitrogenase. The two main peaks are due to Fe–S and Fe–Fe/Mo distances; Reproduced with permission from (a) Hildebrandt, R. L., Andreassen, A. L. & Bauer, S. H. (1970) *J. Phys. Chem.* **74**, 1586 (b) M. R. Antonio *et al.* (1982) *J. Amer. Chem. Soc.* **104**, 4703, copyright American Chemical Society.

Further evidence can therefore be gained by studying the EXAFS of a second type of atom. EXAFS is a particularly important technique because it is not limited to crystals, but is equally effective for amorphous solids, polymers, and liquids — all materials for which structural information is elusive. The example in Fig. 2.5b gives a good illustration of its application to an important but intractable system, but it also indicates the rather limited information given by the technique.

2.3   MOLECULAR ENERGY AND BOND ENERGY

Calculating the strength of particular bonds in a molecule or its overall stability is often the main aim of a study of its bonding. It is therefore very important to be able to check such results against experiment, to confirm the calculations.

The stability of a molecule is measured by its heat (enthalpy) or its free energy of formation from the elements (Box 3). The enthalpy (ΔH_f° measures the strength of the bonds, while the energy (ΔG_f°) describes its overall stability. This information is obtained by standard thermochemical measurements (Cox & Pilcher 1970), and extensive tabulations of values have been produced (e.g. Chase et al. 1986). Despite its fundamental importance, it has become rare for the measurement of heat of formation to form part of the characterization of a new compound. The tabulated values are therefore far from comprehensive, though they cover most simple compounds.

For the bonding of a compound, the heat of formation from $isolated$ $atoms$ (or its reverse, the $heat$ of $atomization$ of the molecule) is more fundamental than the heat of formation from the elements in standard form. The heat of atomization can be calculated with the thermochemical cycle in Fig. 2.6, using the heats of atomization

---

Box 3 Energies

This book uses kilojoules per mole (kJ mol^{-1}) as the principal unit of energy. This allows the comparison of energies at an atomic level with those of reactions. For example, an energy change of 0–10 kJ mol^{-1} represents very slight stability, e.g. of a weak hydrogen bond, and a change of 100–150 kJ mol^{-1} indicates a highly exothermic reaction or a very strong bond. The older unit of kilocalories per mole should, by international agreement, have been abandoned, but regrettably is still often encountered (1 kcal=4.18 kJ).

Those dealing with bonds at the atomic level often concentrate on the energy levels of the individual atom, and measure the energy needed to move an electron from one orbital to another — or to remove it. This can be done (in principle) by applying a voltage. The energy is then measured in electron-volts (eV), the voltage applied to one electron; electron-volts are occasionally used here. The relationship between eV and kJ mol^{-1} is

$$1 \text{ eV} = 96.48 \text{ kJ mol}^{-1} \ (=23.03 \text{ kcal mol}^{-1}).$$

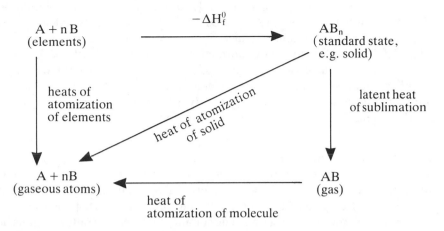

Fig. 2.6 — Thermochemical cycle for the determination of the heat of atomization of a molecule, i.e. its heat of formation from isolated atoms.

of the elements. Whether the heat of atomization of the gas or condensed phase of AB_n is of interest will depend primarily on whether the compound is molecular or extended (e.g. a metal alloy or ionic solid).

2.3.1 Bond energies

In an *extended* solid, the energy of an individual bond is obtained only as a fraction of the total energy for conversion of the compound to its component atoms. However, in a *covalent* molecule it is possible to break one individual bond in the molecule. The energy to do this is known as the *bond dissociation energy* (BDE) for that bond. Generally, the BDE is less well-defined and more difficult to measure than the overall stability, though it is often of more interest. In the simplest case, a diatomic molecule, the BDE is just the dissociation energy of the molecule. With a simple molecule, such as CH_4, the BDE for each bond is also easy to determine. All four C–H bonds are identical and so the individual BDEs must be 0.25 times the total dissociation energy to atoms ($415 = 0.25 \times 1660$ kJ mol^{-1}).

Unfortunately, in a molecule with more than one type of bond, it is impossible to decide how the total dissociation energy should be divided between the individual bonds. It is also impossible to measure directly the BDE for one bond in a many-atom molecule. It is often difficult in practice to dissociate *one* particular atom from the molecule and determine the energy needed to remove it. Even if this is done, the measured energy is not the BDE for that bond. If one C–H bond in CH_4 is broken, the $\cdot CH_3$ radical produced does not have the energy or electronic arrangement of three individual C–H bonds. In addition, the pyramidal $\cdot CH_3$ radical first produced rearranges to a planar form of lower energy. Thus the energy for the process

$$CH_4 \rightarrow H\cdot + \cdot CH_3$$

with all species in their lowest energy state is different again.

The resolution of this problem depends on the assumption that similar bonds in different compounds have the same BDE. It is then possible to calculate unknown BDEs by using these assumed values. For example, the BDE for the C–C bond in C_2H_6, 329 kJ mol^{-1}, is found by subtracting 6×415 kJ mol^{-1} (the BDE for C–H in CH_4) from the total dissociation energy. This method can be widely applied if suitable compounds are chosen. It produces results that are broadly consistent from compound to compound, but the underlying assumption cannot be completely tested; Dasent (1970) lists BDE values for a number of elements.

2.3.2   Conformational energy

In a flexible molecule, the relative energies of the alternative conformations are important in deciding what will be its preferred shape. This will control, for example, whether it will be bound by a particular enzyme, and perhaps act as a drug by interfering with the enzyme action. The stability of different conformations depends partly on the energy consequences of distorting bond lengths and angles from their ideal values, and partly on the attractions and repulsions between non-bonded atoms.

Some of the distortion energies can be measured by infrared spectroscopy, and the non-bonded interactions can be calculated by quantum mechanics. However, an empirical approach has been found most useful, in which the parameters are adjusted so that observed conformations and their energies are reproduced by calculation. These parameters are then applied in calculating the conformations of unknown molecules. These calculations are called *molecular mechanics* or *molecular dynamics*, depending on whether they determine the equilibrium conformation of a molecule, or the energetics of a reaction leading from one state to another. For organic molecules, these methods have been remarkably successful in calculating geometries that agree well with experimental observations. They have been extended to coordination complexes of metals with rather less success, because of difficulties in obtaining satisfactory parameters for the interactions involving the metal atoms.

2.3.3   Distance/bond energy correlations

An entirely different approach to bond strength is based on the belief that the shorter the bond between two atoms, the stronger it is. Indeed, this provides a large part of the interest in determining structures to a greater precision than that needed for identifying connectivity. At first sight it is not clear what use can be made of the information on the Re–Re and B–B bond lengths shown in Fig. 2.7. Two assumptions underlie the application of such evidence.

(i)   The distance between two atoms is related to the strength of the bond (strictly the *force*) between them. The nature of this relationship is considered in section 5.8.

(ii) It is possible to determine radii that indicate the size of an atom in a particular type of bond, and these radii can be transferred from one compound to another. This feature of *self-consistency* is crucial. Clearly, radii are of very little use if they apply only to the single compound from which they have been determined.

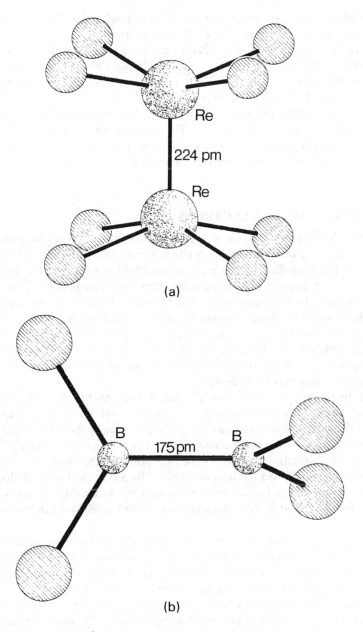

Fig. 2.7 — (a) $Re_2Cl_8^{2-}$ Re–Re distance 224pm (b) B_2Cl_4 B–B distance 175pm.

The radii depend on the type of bonding between the atoms. Thus, an atom may have characteristic *metallic, ionic,* or *covalent* radii. The calculation of these radii is not always easy, and the problems are examined in Chapters 3 to 5. As an illustration of their application, we can consider the two examples just given. For B, a standard single *covalent* radius can be estimated as 83.0 pm, while for Re, the *metallic* radius (in the element) is 137.5 pm. The B–B bond is significantly longer than twice the covalent radius, indicating that this is rather a weak single bond. In contrast the Re-Re distance is 81% of that in the metal, implying a very strong bond. In fact, a detailed study leads to its description as a *quadruple* bond — the strongest type of bond yet observed.

Ionic and metallic distances show much less variation from compound to compound than do covalent distances. The corresponding radii are therefore most useful in considering compounds in which two or more metals or ions are present, to identify the way in which the atoms can pack together; this is explored in Part III.

2.4 DISTRIBUTION OF ELECTRONS

The distribution of electrons in a molecule is one of the main predictions of bonding theories. Checking this with experiment is even more difficult than with energy predictions. Electron distribution may be described at a simple level in terms of the orbitals occupied. In more detail, the point-by-point electron density for each orbital may be calculated by quantum mechanical methods. Details of these calculations are outside the scope of this book, though it is worth noting that they can be performed at various levels of sophistication. It is in principle possible to calculate the electron density accurately for molecules of any complexity. However, if some of the atoms have large numbers of electrons, or if very precise results are needed, inordinately long computer runs may be necessary.

From the information in Table 2.2, that X-rays interact with the electrons in a compound, we would expect X-ray diffraction to reveal the electron density directly. We might therefore expect to get a clear picture of the electrons involved in bonding. Unfortunately this overlooks one vital fact. X-ray diffraction shows *all* the electrons and makes no distinction between core and bonding electrons. Furthermore, the nuclear charge on each atom is best neutralized by a spherical shell of electrons. As a consequence, the overall electron density seen by X-ray diffraction is very nearly spherical about each atom; this is clearly visible in Fig. 2.4a.† Fortunately all is not lost!

With very careful measurement of the X-ray diffraction intensities, it is possible to calculate *difference electron density maps*. These show the differences between spherical atoms and the actual electron density, and they do reveal features which relate to valence electrons and bonding. The example in Fig. 2.4d shows three such features: small positive peaks in the centre of each bond, corresponding to the electrons taking part in the bond; small negative peaks elsewhere around the atoms showing from where the bonding electrons have moved; a rather larger positive peak corresponding to the lone pair on the oxygen atom.‡

---

† In addition, the vibrations of the individual atoms caused by their thermal energy smear out the almost spherical electron density, usually into an ellipsoid. For example, the O atom in a C=O group will vibrate more from side to side than along the bond direction. These so-called *anisotropic thermal vibrations* produce much greater effects than does asymmetry in the electron distribution due to bonding.

‡ The one very specialized circumstance in which we can see the distribution of an individual electron directly is mentioned above: neutron diffraction measurement of the magnetic scattering from a paramagnetic material, as shown in Fig. 2.4c.

2.4.1   Orbital distribution and electron energy

The peaks on difference electron density maps provide no information about the orbitals which the electrons occupy, nor about their binding energy. However, the maps can be matched with theoretical calculations, to give a rather indirect check on the latter. More satisfactory confirmation comes from direct observation of electron binding energy, by the technique of *photo-electron spectroscopy*. This uses a beam of photons with sufficient energy (E_1) to cause the ejection of electrons from the orbitals of the compound concerned (usually u-v radiation). The energies of the electrons produced are measured (E_2), and their binding energy (BE) is given by

$$\mathrm{BE} = E_1 - E_2$$

Photoelectron spectra are complicated by vibrational effects, so that determination of the precise energy of the ground state may be more difficult than this simple description would suggest. Examples of photoelectron spectra are given in Chapter 5, showing how they can be matched with molecular orbital diagrams.

2.5   APPENDIX: THE CRYSTAL LATTICE

The lattice in a crystal is a purely mathematical concept which underlies the identification of a unit cell. It can be defined as:

The array of all points in a crystal which have identical environment

The choice of *one* lattice point is entirely arbitrary, though it may be convenient to pick a point with a special relationship to whatever symmetry elements the crystal contains (e.g. on a 2-fold axis of rotation). However, once one point is picked, all the remaining lattice points are fixed according to the definition. This is most easily illustrated in two dimensions (Fig. 2.8). The point X has been selected, and all the other lattice points are marked with dots.

Once the lattice points have been identified, a unit cell (the unit whose repetition generates the crystal) is selected that is consistent with them. The unit cell is *any parallelepiped* (or in two dimensions a parallelogram) *which has lattice points at each corner*. In Fig. 2.8, two choices of unit cell show how their repetition produces the whole crystal.

Unit cell II (Fig. 2.9b) differs from I (Fig. 2.9a) in that it contains a lattice point at its centre, as well as at each corner. Such a unit cell is called *centred*, in contrast to unit cell I which is *primitive*. It is obvious from Fig. 2.8 that any centred cell can be redrawn as a primitive unit cell, e.g. II as I, and repetition of either primitive or centred cell generates the complete crystal. However, if the crystal contains elements of symmetry, this redrawing may destroy the relationship between the unit cell and the symmetry. Again, this is illustrated in two dimensions (Fig. 2.10). This lattice has two planes of mirror symmetry (down and across the page). The first lattice point has been chosen to lie on the intersection of these planes. Related planes pass through every other lattice point. The simplest choice of a unit cell whose axes lie along the mirror planes is I, which is *centred*. Although primitive unit cells can be chosen (e.g. II and III), they do not reflect the two planes of symmetry.

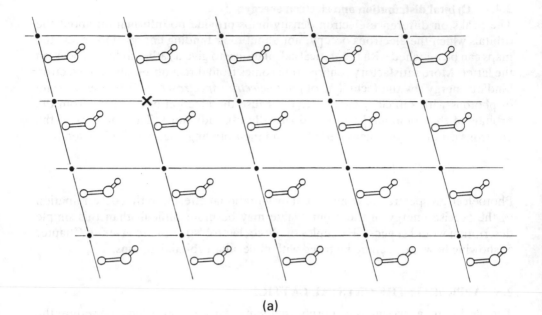

(a)

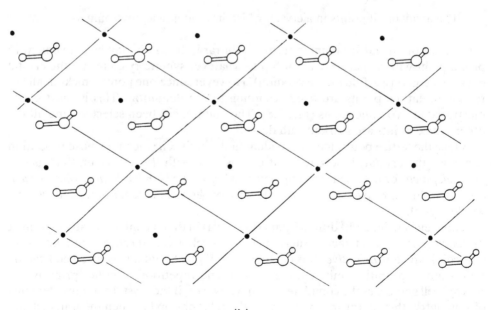

(b)

Fig. 2.8 — Lattice points in two dimensions. (a) One lattice point (X) is arbitrarily chosen, and
this defines the positions of the remaining lattice points (dots). One choice (I) of a primitive unit
cell is shown. (b) A second choice (II) of unit cell. This is *non-primitive*, i.e. contains lattice
points within the cell as well as at the corners.

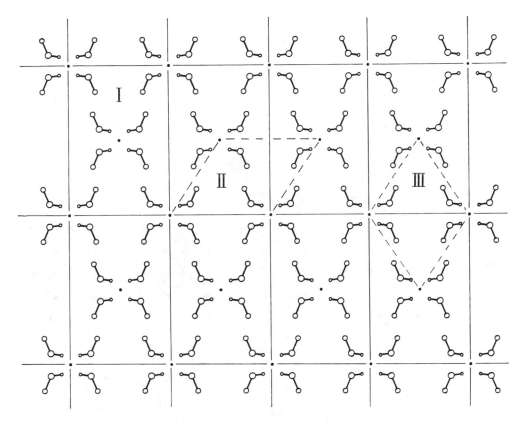

Fig. 2.9 — A centred unit cell (I) containing planes of mirror symmetry. Two possible primitive unit cells (II, III; dashed) using the same lattice points are valid choices, but lose the relationship between the cell and the two mirror plane directions.

Exactly the same applies in three dimensions. Lattice points can appear in the centre of *one* face of the unit cell (as in cell I in Fig. 2.9), or in the centres of *all three* faces (a 'face-centred' cell), but not of two faces.† A centring lattice point may also exist in the middle of the unit cell (a 'body-centred' cell). Bravais in 1848 identified the 14 arrangements of lattice points (distributed among the 7 crystal systems) which cannot be simplified without losing the symmetry of the crystal system. These are the Bravais lattices, which are drawn in Fig. 2.10.

† The proof of the non-existence of lattices centred in both A and B faces can be found in crystallographic textbooks, e.g. Sands (1960).

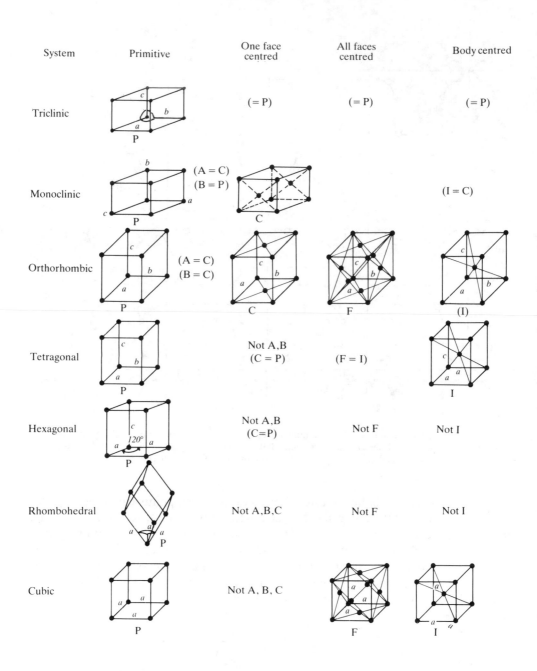

Fig. 2.10 — The Bravais lattices. The diagrams indicate the form of the different crystal lattices, and the table shows their systematic relationships. Note: In the monoclinic and orthorhombic systems, the Bravais lattices with one face centred are conventionally labelled as C-centred.

General surveys

Ebsworth, E. A. V., Rankin, D. W. H. & Cradock, S. (1987) *Structural methods in inorganic chemistry*. Blackwell Scientific, Oxford.

Cheetham, A. K. & Day, P. (1986) *Solid state chemistry techniques*. Clarendon Press, Oxford.

Diffraction

Glusker, J. P. & Trueblood, K. (1985) *Crystal structure analysis*. Oxford University Press, New York (2nd ed.).

For standard bond distances, see Allen, F. H., Kennard, O., Watson, D. G., Brammer, L., Orpen, A. G., & Taylor, R. (1987) *J. Chem. Soc. Perkin II* p. S1; Orpen, A. G., Brammer, L., Allen, F. H., Kennard, O., Watson, D. G. & Taylor, R. (1989) *J. Chem. Soc. Dalton* p. S1.

EXAFS

Koningsberger, D. C. & Prins, R. (eds) (1988) *X-ray absorption, principles, applications, techniques of EXAFS, SEXAFS and XANES*. Wiley, New York.

Photoelectron spectroscopy

Brundle, C. R. & Baker, A. D. (eds) (1977) *Electron spectroscopy: theory, techniques and applications*. Academic Press, London. Vol. I.

Thermochemistry

Chase, M. W. *et al.* (1986) *JANAF Thermochemical tables*. American Chemical Society, Michigan.

Cox, J. D. & Pilcher, G. (1970) *Thermochemistry of organic and organometallic compounds*. Academic Press, New York.

Dasent, W. E. (1970) *Inorganic energetics*. Penguin, Harmondsworth.

Molecular mechanics

Burkert, U. & Allinger, N. L. (1982) *Molecular mechanics*. American Chemical Soc., Michigan. Monograph 177.

Theoretical calculations

Clark, T. (1985) *A handbook of computational chemistry*. Wiley, New York.

Crystal lattices

Sands, D. F. (1960) *Introduction to crystallography*. Benjamin, New York.

Steadman, R. (1982) *Crystallography*. Van Nostrand Reinhold, Wokingham, UK.

Text reference

Jones, P. G. & Lautner, J. (1989) *Acta Cryst.* C44, 2089.

Part II
Ideal bonds

3

Metals

3.1 INTRODUCTION

Elements which attract electrons weakly, i.e. are of low electronegativity, and materials containing two (or more) such elements lie in the top left corner of the bond triangle and show metallic behaviour (Fig. 3.1). In the simplest model of metallic

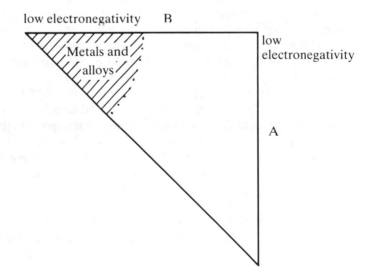

low electronegativity   B

Metals and
alloys

low
electronegativity

A

Fig. 3.1 — Metals in the bond triangle.

bonding, the metal atoms are assumed to lose their outer electrons completely. Their positive charges then provide a uniform attraction for the electrons. The attraction ends at the surface of the metal, so the positive potential has the form shown in Fig.

3.2. Because the electrons are free from any localized attraction to the metal centres, this model is called the *free electron model.*

3.2 STRUCTURES OF METALS

In predicting the structures of metals by using this bonding model, we need to consider both attractive and repulsive forces. As the metal atoms get closer together, the positive potential increases and the electrons are more strongly held. However, the core electrons of each atom are not involved in metallic bonding. This means that as two atoms approach each other, these core electrons start to repel each other, and this repulsion increases very rapidly as they get close together (see section 4.7.2). The simplest description of this behaviour uses the *hard-sphere* model. The atoms are treated as hard spheres which cannot be closer together than a definite minimum distance; the repulsions are ignored if the spheres are not in contact.

3.2.1 Close-packed structures

Metals are therefore predicted to have structures in which these hard spheres are packed as closely together as possible, giving maximum binding of the electrons. To work out the arrangement of metal atoms in such a *close-packed* structure, consider one layer of spheres.† An obvious arrangement is the square array shown in Fig. 3.3a. However, the staggered arrangement in Fig. 3.3b is more effective (as long as the layer extends for a fairly large distance). Each sphere in (b) touches six other spheres in the layer (as shown for the sphere labelled A). The same principle applies when we stack the layers together; it is more effective to place the atoms of the next layer in the hollows between the atoms of the first layer rather than directly on top of them. The best position for stacking can be seen in Fig. 3.4. Each sphere fits into the hollow between three spheres in the layer below (Fig. 3.4a).

The extended view (Fig. 3.4b) shows that the next layer only fills up only *half* the hollows in the first layer. This is important when a third layer is added. The second layer contains two types of hollows. Those marked (y) correspond to the positions of *atoms* in the first layer, while those marked (z) lie over *holes* in the first layer. This leads to two ways of extending the stack of close-packed layers. The next layer can fit either in the (y) or the (z) hollows. Conventionally, these are distinguished by labelling the layers according to their position. A is used for the first layer and B for the second. The first arrangement has the third layer in the position (y), so the sequence is ABA. Alternatively, the third layer is different from either of the others (z) and is labelled C, giving an ABC sequence. Both sequences can be continued indefinitely

..ABABAB..   or   ..ABCABCABC..

† It is far easier to understand the packing of spheres from models than from pictures, and readers not already familiar with the structures of metals should if possible use models while studying this section. Various commercial models are available but it is also easy to construct layers of spheres from table-tennis balls stuck together with polystyrene cement. It is useful to have two or three layers with about ten spheres and two or three with three spheres.

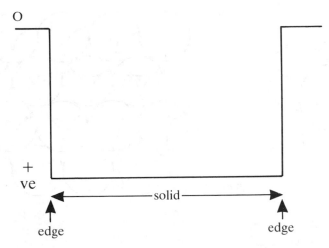

Fig. 3.2 — Variation of positive potential in the simplest model of a metal. The depth of the potential well depends on the charge on each metal ion and on their spacing.

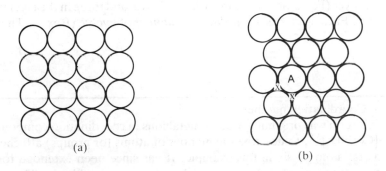

Fig. 3.3 — Packing of spheres in layers (a) square array, (b) closest-packed array, in which the same number of spheres take up less area than in (a); atom A has six neighbours in this layer. The possible possible positions (X) for atoms in the next layer are marked.

Both represent *close-packed structures*, and are equally effective ways of achieving the best possible packing of spheres.†

Two key characteristics are shared by both structures. Firstly, the atoms occupy 74% of the available space, which can be shown to be the maximum possible for packing equal-sized spheres. Secondly, each atom is in contact with twelve others: six

† Metals occasionally take up more complicated variants of these stackings, such as a repetition of the layer sequence . . (ABAC) . . or . . {ABABC} . .

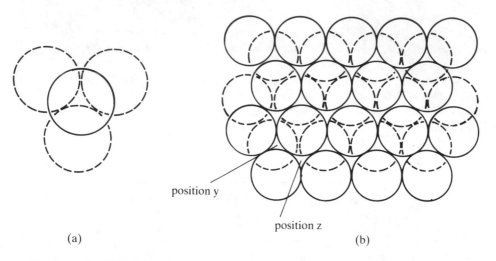

position y

position z

(a)   (b)

Fig. 3.4 — Stacking of layers of spheres (a) position of one sphere over a layer of spheres (dotted), (b) overall stacking of *two* layers (lower layer dotted). The possible positions for one atom of a third layer are marked. Position (y) gives sequence ABA with third layer over the first layer. Position (z) gives sequence ABC with the third layer not over the first layer.

in the same layer (Fig. 3.3b), three in the layer above, and three in the layer below (as in Fig. 3.4a). Each atom has a *coordination number* of twelve (Box 4). The metallic

---

Box 4 Coordination number

This term was originally used of metal ions in coordination compounds, e.g. $[Co(NH_3]^{3+}$, to describe the number of atoms (or groups) attached to the metal atom — six in this example. It has since been extended to any atom, as a convenient term for the number of nearest neighbours. The bonds may be:

covalent	CH_4	C has coordination number 4
ionic	NaCl	Na and Cl have coordination number 6
metallic	Mg	Mg has coordination number 12

Problems arise when there is no clear distinction between the nearest neighbours and those rather further away (e.g. in the body-centred cubic structure, section 3.2.2). For covalent compounds, the coordination number can be taken as the number of atoms with bonds to the central atom, e.g. four in $(HO)_3P{=}O$, despite there being longer and shorter P–O bonds. Some metallic and ionic compounds show real ambiguity (e.g. LaF_3 with seven F at 242–249 pm, one at 264 pm, and one at 300 pm). In discussing these structures, it is useless to insist on a rigid definition of coordination number.

radii in these structures can easily be obtained, by taking half of the M–M distance. Standard metallic radii (tabulated in Appendix A.3) have been derived for 12-coordinate metals, though there has been no detailed study of the consistency of metallic radii from one structure to another. They are particularly useful in understanding the ways in which different metals pack in alloys (see section 6.4.3).

The two close-packed structures are named according to the crystal systems (section 2.1.1) to which they belong as either *hexagonal close-packed*, often abbreviated to h.c.p. (ABAB sequence), or *cubic close-packed*, c.c.p. (ABCABC sequence). Figs 3.5–3.6 show conventional and three-dimensional representations of each structure. It is reasonably easy to see the close-packed layers in Fig. 3.5a, considering for example the atoms at height 0, but their identification in the conventional representation of c.c.p. (Fig. 3.5b) is very difficult. In fact, the layers are perpendicular to the body diagonal of the cube. Fig. 3.6c–d shows how they are revealed by removing atoms from one corner. Fig. 3.6 also shows that each face of the cube has an atom at its centre. Because of this and for contrast with the next structure to be described, the cubic close-packed structure is sometimes named 'face-centred cubic', from its Bravais lattice (section 2.5). This alternative name is less informative than 'cubic close-packed', and is not recommended.

3.2.2   The body-centred cubic structure

The majority of metals take up either the h.c.p. or c.c.p structures, but one other structure is surprisingly common. This is the *body-centred cubic* or b.c.c structure (named from its Bravais lattice!). The basic structure consists of a cube with a metal atom at each corner, and one atom in the centre (Fig. 3.7). This atom obviously has a coordination number of eight. If the structure is extended, we see that the corner atoms are also eight-coordinate. Looking at the adjacent unit cells, we see that the middle atom has six *next*-nearest neighbours, the central atoms in each adjoining cell. These are only slightly further away than the eight nearest neighbours (1.15:1.0 ratio of distances), and the b.c.c. structure is sometimes loosely described as containing 14-coordinate metal atoms.

Only 68% of space is occupied by atoms in the b.c.c structure, compared to 74% for the close-packed structures. According to our original criterion for metallic packing, that they will be as dense as possible for a given size of atom, we would expect the b.c.c. structure to be strongly disfavoured because of its lower packing density. Unexpectedly, this argument is not correct. Several metals show both b.c.c. and a close-packed structure. At the phase transition between the two structures, the metallic radii can be determined for both structures. It is found that the 12-coordinate atoms in the close-packed structure are *larger* than the 8-coordinate atoms in the body-centred cubic structure. The ratio between the metallic radii varies only slightly from example to example, with an average value of 1.028:1. This corresponds to a change in the volume of each atom of $(1.028)^3 = 1.086$ — the same as the ratio between their packing densities (1.088) (Pearson 1985). Thus this size change compensates exactly for the less effective packing of the b.c.c. structure. This behaviour shows that the hard-sphere model of metal atoms does not give a perfect description of metals, but that their radii are affected by changes in coordination number. We can understand this from the forces acting on an individual metal atom. The total bonding energy does not vary on going from the 8- to the 12-coordinate

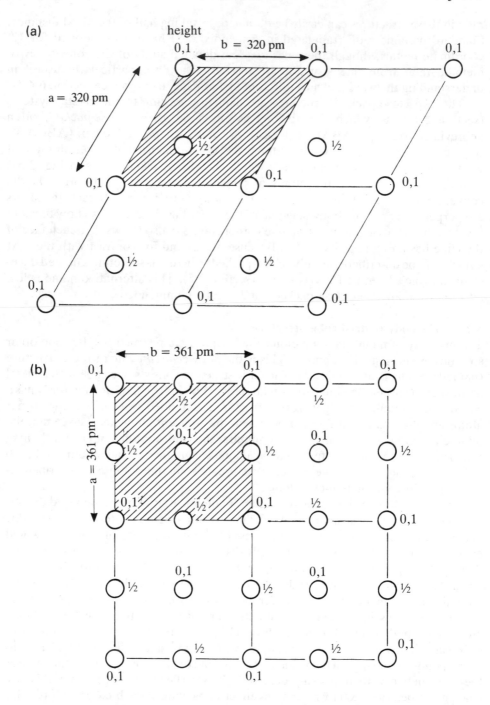

Fig. 3.5 — Conventional representations of close-packed metals (a) Mg metal, with the h.c.p. structure, showing four unit cells (with one unit cell shaded). The height of each atom in fractional coordinates (Box 2) is marked against it. The c-axis (perpendicular to the paper) is 520 pm long. (b) Cu, with the c.c.p. structure; the c-axis is 361 pm long.

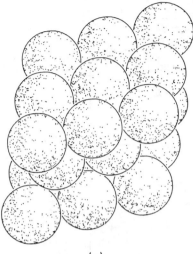
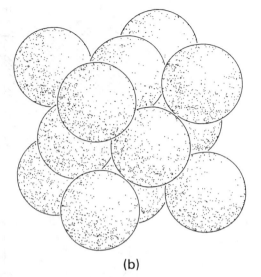

(a)   (b)

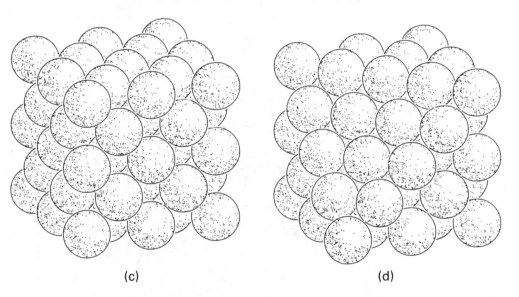

(c)   (d)

Fig. 3.6 — (a) Four unit cells of a h.c.p. metal, (b) one unit cell of a cubic close-packed metal, (c) eight unit cells, (d) as (c) with some corner atoms removed to show the close-packed layers.

structure, but for the former, the corresponding force is exerted through only eight contacts. This larger force can be expected to compress the atom more.

The choice by a particular metal of the h.c.p., c.c.p., or b.c.c. structures is often related to the number of valence electrons and their energies (see section 6.2). It has also been found that the b.c.c. structure appears as a high-temperature phase for

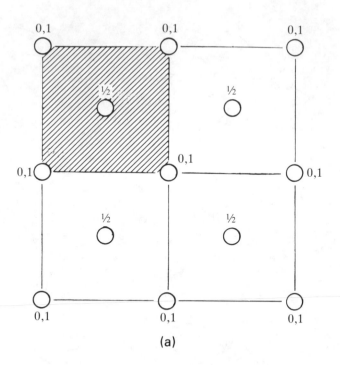

(a)

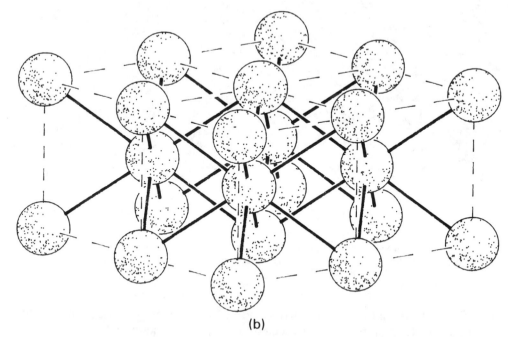

(b)

Fig. 3.7 — The body-centred cubic structure (a) conventional representation showing four unit cells, (b) view of four unit cells, showing the eight neighbours of an atom at the centre of each cell. The atoms at the corner of the cell also have eight neighbours; the outlined cubes can be moved so that the central atoms are at the corners of the cells, without altering the structure.

metals which are close-packed at lower temperatures. Thus, Ca and Sr convert from h.c.p. to b.c.c. at 450°C and 621°C respectively, while Li and Na become close-packed on reducing the temperature to 78 K and 36 K respectively. K, Rb, and Cs are b.c.c. at all temperatures — perhaps because their transition temperatures are below absolute zero!

In these simple metals, the stabilization of b.c.c. at high temperature can be seen as a consequence of the increasing amplitude of the atomic vibrations as the temperature rises. The presence of eight rather than twelve nearest neighbours allows the metal atom more space to vibrate, avoiding the repulsion of adjacent atoms. This suggestion is supported by the behaviour of the coefficient of expansion of metallic calcium. This coefficient reflects the forces exerted by each atom through its vibrations: the larger the coefficient, the greater the force being transmitted to the adjoining atoms, causing the crystal to expand. At the transition temperature, the coefficient of expansion is 20% less for the b.c.c. phase of Ca than for its h.c.p. phase.

We can now understand why the b.c.c. phase appears, if we assume that the metallic bond energy depends on the density (as shown experimentally in section 6.3). Consider a metal for which the close-packed phase is the more dense at low temperature. As the temperature rises it expands more than the (unstable) b.c.c. phase. Thus, the transition occurs at precisely the temperature at which the densities are equal, as observed experimentally. Above this temperature, the b.c.c. phase is the more dense and so it is stable.

3.3   ENERGY OF SIMPLE METALS

Using the 'free electron' model for metallic bonding described by the potential shown in Fig. 3.1, we can understand much of the character of metals and can even calculate values for some of their properties. According to the quantum theory of matter, electrons in such a potential field (or any other) are restricted to specific energies — the quantum levels. Each level is identified by *three* quantum numbers.†
The first step is to discover these energies. This is achieved by solving the *Schröd-inger equation* for this system. Although this is mathematically not difficult, it requires the use of differential equations and is relegated to the Appendix to this chapter.

From the solutions of the Schrödinger equation we obtain the energy of the electron for *each* permitted energy level; in this case, the energies are a simple function of the quantum numbers. We also obtain the actual electron wave function ψ, though this is less directly useful. For a cubical metal crystal, these energies are given by:

$$E = (n_x^2 + n_y^2 + n_z^2)h^2/8ma^2 \tag{3.1}$$

(m = mass of electron; h = Planck's constant; a = dimension of cube).

† There are three quantum numbers because we are considering a three-dimensional structure. It is sometimes useful for simplicity to consider the one-dimensional case, which has only one quantum number.

The vital components of equation (3.1) are the integers n_x, n_y, n_z. These are the quantum numbers which identify each possible wave function and corresponding electron energy. According to the Pauli principle, each energy level holds 2 electrons (of opposite spin). Clearly, an enormous number of energy levels are needed for the electrons of (say) 1 mole (23 gm) of Na. This contains 6×10^{23} atoms, each contributing one electron to the metal, occupying 3×10^{23} energy levels.

To apply equation (3.1), say to metallic sodium, we need to work out the maximum energy of these levels. This can be done by a small manipulation of the equation. The equation for a sphere of radius r is given by

$$r^2 = x^2 + y^2 + z^2 \tag{3.2}$$

We can therefore treat the $(n_x^2 + n_y^2 + n_z^2)$ term of (3.1) as defining a sphere (or a circle in two dimensions; Fig. 3.8). Each set of values of n_x, n_y, n_z (one energy level)

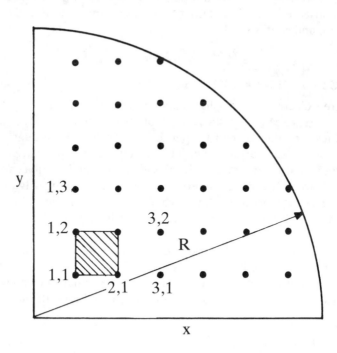

Fig. 3.8 — Quantum numbers (n_x, n_y), for electrons in a metal with energy up to E_{max}, in two dimensions lie within a circle $(E_{max} \sim R^2 = n_x^2 + n_y^2)$. Each combination (n_x, n_y) is associated with a unit square (e.g. $n_x = n_y = 1$ with the shaded square). Thus the number of energy level is the *area* of the quadrant of the circle of radius R. Similarly, in three dimensions, each combination (n_x, n_y, n_z) is associated with a unit cube, and the number of energy levels is the *volume* of one octant of the sphere of radius R.

corresponds to one point within the sphere, and is associated with one unit cube. The total number of values within a sphere of radius r is just the number of unit cubes, i.e. the volume of that sphere. Because n_x, n_y, n_z can only be positive, we consider only one octant (1/8th) of the whole sphere. Then

$$\text{number of levels} = 1/8.4/3.\pi.r^3 = 1/6.\pi.r^3 \tag{3.3}$$

From (3.1) and (3.2)

$$n_x^2 + n_y^2 + n_z^2 \;(=r^2) = E.8ma^2/h^2 \tag{3.4}$$

Each level holds 2 electrons, so N, the number of electrons up to an energy E_{max}, is given by

$$N = 1/3\pi r^3 = 1/3\pi (8ma^2 E_{max}/h^2)^{3/2} \tag{3.5}$$

$$= 8/3\pi a^3 (2mE_{max}/h^2)^{3/2} \tag{3.6}$$

Usually, this is related to unit volume by taking $a=1$. Then

$$E_{max} = (3N/8\pi)^{2/3}.(h^2/2m) \tag{3.7}$$

where N is the number of electrons in this volume.

By inserting standard constants, equation (3.7) gives E_{max} in electron volts as

$$E_{max} = 3.61.10^5 \, (nA/V)^{2/3} \tag{3.8}$$

where n = number of electrons released by each atom, V = volume of unit cell (pm^3), A = number of atoms in each unit cell, so that (nA/V) is the density of electrons in the metal (el. pm^{-3}). For Na with $A=2$ and $a=423.5\,pm$ for its b.c.c. cell, $E_{max} = 3.2\,eV$ ($310\,kJ\,mol^{-1}$). E_{max} can also be measured experimentally by soft X-ray emission (Fig. 3.9). An electron is excited from one of the core levels (1s, 2s, or 2p) of an Na atom; the empty orbital is then refilled by one of the metallic electrons, and a quantum of X-rays is emitted. This varies in energy, depending on whether the refilling electron comes from the top or the bottom of the band. The energy range for the emission corresponds precisely to the energy ($0-E_{max}$) of the metallic electrons. For Na, the experimental value is $2.8\,eV$, in reasonable agreement with the very simple theory. However, for the transition metals the calculated value is much further from experiment, and the theoretical approach needs to be modified (Chapter 6).

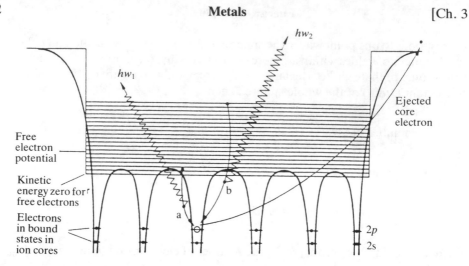

Fig. 3.9 — The measurement of E_{max} by soft X-ray emission. The metallic energy levels are indicated by horizontal lines, and the localized 1s, 2s, and 2p energy levels on individual atoms are at the bottom of the diagram. Reproduced with permission from Coles, B. R. & Caplin, A. D. (1976) *Electronic structure of solids*. Edward Arnold, London, p. 51.

As well as the formula for N (the number of electrons for which orbitals are available up to a particular energy E_{max}) a related quantity is particularly useful: $N(E)$, the number of energy levels with a particular energy E, i.e. the *density* of energy levels. This is simply the derivative of N. Thus, from equation (3.6)

$$dN/dE = N(E) = 4\pi(2m/h^2)^{3/2}.E^{1/2} \tag{3.9}$$

The relationship between $N(E)$ and E is a parabola, which is usually drawn out with E on the horizontal axis (Fig. 3.10). All the energy levels in the shaded area up to E_{max} are occupied. The term *band* is used for a group of energy levels, such as those in Fig. 3.10, which are so close in energy that they can be treated as continuous.

3.4   PROPERTIES OF METALS

3.4.1   Ductility

Using the free-electron model, many properties of metals become understandable. We look here at some examples. In qualitative terms, one of the simplest is the *ductility*, the ability to deform (the opposite of brittleness). If under stress, one metal atom is displaced, it will form equally strong bonds in its new environment, because the metallic bond depends only on the closeness of the individual atoms. However, the *strengths* of metals are very variable, from the softest alkali metals to the hardest transition metal alloys, and the factors controlling this property are more complex. They will be examined in Chapter 6.

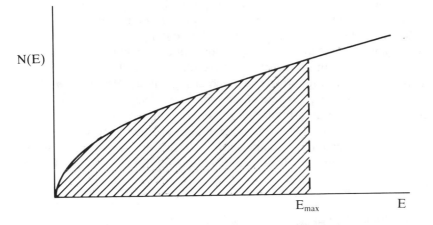

Fig. 3.10 — Relationship of E, the metallic electron energy, and $N(E)$, the *density* of energy levels with energy E (number of levels per unit volume). The shaded levels are occupied for the value of E_{max} shown.

3.4.2 Thermal and electrical conductivity

The high thermal and electrical conductivity of metals is a direct consequence of their electronic structure. Both depend on the presence of empty orbitals lying just above those occupied. Electrons can easily be promoted into these orbitals. An electron excited into an empty orbital at one end of the crystal (by an electric potential or by external heat energy) can then move freely to the other end of the crystal. Remarkably, because electrons are involved in both types of conductivity, the ratio

$$L \text{ (the Lorenz number)} = K/\sigma T$$

where K = thermal conductivity $(Jm^{-1}K^{-1}s^{-1})$
   σ = electrical conductivity $(ohm^{-1}m^{-1})$
   T = temperature (K)

should be the same for all metals.

$$L = 1/3(\pi k/e)^2 = 2.45 \times 10^{-8} \text{ watt ohm K}^{-2}$$

(k = Boltzmann's constant, e = charge of an electron). The measured values of L for most metals are indeed close to this, e.g. 2.12×10^{-8} for Na, 2.31×10^{-8} for Ag.

3.4.3 Paramagnetism

The magnetic properties of metals pose more problems. Apparently all the metal orbitals are occupied by paired electrons, and so the metal should be diamagnetic

(Box 5). However, a magnetic field has a rather different effect on electrons in metallic orbitals than it does on the electrons in a covalent diamagnetic solid. Fig. 3.11a is the same as Fig. 3.10, except that the top half of the diagram shows the spin-up electrons and the bottom half the spin-down ones. If a magnetic field is applied, parallel (say) to the spin-up direction, these electrons will become slightly more stable (Fig. 3.11b, with the difference much exaggerated). Because empty orbitals are available just above E_{max}, some of the spin-down electrons can reverse direction and move to empty spin-up orbitals of lower energy (Fig. 3.11c). The slight excess of spin-up electrons then produces paramagnetic behaviour. The number of excess electrons is not affected by temperature, so this is *temperature independent paramagnetism*.† The formula for the magnetic susceptibility of a metal is remarkably simple

$$\chi = 2 . \mu_0 . \mu_B . N(E_{max})$$

where μ_0 = magnetic permeability of a vacuum, μ_B = Bohr magneton.

The effect of an electric field is very similar to that of a magnetic field. It decreases the energy of electrons moving in the direction of the field, compared to those moving in the opposite direction. The result is a net excess of electrons moving with the field, i.e. electrical conduction.

3.4.4   Optical properties
The opacity and reflectivity of metals are also consequences of their delocalized electron clouds, but the precise relationship is far from straightforward. The continuity between filled and empty orbitals in the metal should mean that light of any frequency can be absorbed. This is true for frequencies up to and including that of visible light. At the same time, most of the light is immediately re-emitted, causing the metal to be reflective. However, the incident radiation interacts strongly with the electrons in the metal, which oscillate in response to the fluctuating electromagnetic field. Above a certain frequency, known as the *plasma* frequency (which depends on the metal being examined), the electrons cease to respond and the metal becomes transparent. This frequency lies in the ultraviolet region for all metals, but for a few, including Cu and Au, it is quite near the visible. Their reflection of blue light is not complete, and they appear red or yellow. Some materials, which show metallic conductivity but have much lower concentrations of metallic electrons, have plasma frequencies in the visible or infrared region. An example is SnO_2 doped with 1–3% Sb, which is transparent to visible light, having its plasma frequency in the near infrared (Cox 1987, p. 42).

3.4.5   Specific heat of metals
The final property of metals to be discussed is the calculation of their specific heat. In the early years of this century, this represented one of the major problems in physics.

---

† This contrasts with the magnetic behaviour of unpaired electrons in transition metal complexes. There, the alignment of individual electrons parallel to the magnetic field is opposed by their thermal agitation. In metals, the electron energy bands are shifted as a whole.

Box 5 Magnetic behaviour

The terms paramagnetism and diamagnetism describe the behaviour of a material in a magnetic field.

Diamagnetic — material weakly repelled from a magnetic field interacting with *pairs* of electrons (Fig. B5(a)). Paramagnetic — material rather strongly attracted to a magnetic field. This is caused by the magnetic field interacting with *unpaired* electrons, and is a valuable test for the presence (and the number) of these electrons (Fig. B5(b)). In most paramagnetic materials, the orientation of the unpaired electrons by the magnetic field is opposed by thermal agitation, so the observed paramagnetism increases as the temperature decreases. In some materials, the paramagnetism is not affected by temperature. This behaviour is called *temperature-independent paramagnetism*.

(a)
diamagnetic
material

(b)
paramagnetic
material

Fig. B5 — Effect of a magnetic field on (a) a diamagnetic material (small force) (b) a paramagnetic material (large force).

Other types of magnetism are found: ferro- and anti-ferromagnetism are caused by unpaired paramagnetic electrons acting together (section 6.3.2). In the case of ferromagnetism, the individual electrons are aligned parallel to each other, so that the material has an overall magnetic moment. In anti-ferromagnetism, the electrons point in directions that cancel out each other's moments, so that the overall magnetic moment is zero, even though each magnetic atom has a non-zero moment. In ferrimagnetism, which is found in magnetic materials like iron, electron spins point both up and down, but in unequal numbers, so that there is a net magnetic effect.

The thermal and electrical properties of metals indicate the availability of large numbers of electrons. These should all be able to absorb energy, leading to a very high specific heat. Instead, experiments showed that the specific heat was almost the same as that expected for the metal atoms alone, with only a 1% discrepancy (at room temperature) which could be attributed to the electrons. This electronic contribution can be measured accurately by extrapolating the specific heat determinations to 0 K, when the atomic contribution ceases to dominate.

To resolve the problem, we need to consider the precise effect of thermal excitation on the electron energies. The amount of energy available to each electron is of the order of kT (0.026 eV, 2.5 kJ mol^{-1} at 300 K). This is far less than E_{max}, the maximum energy of the band of metallic electrons (2.8 eV for Na, as noted above). Supplying kT units of energy to an electron low down in the band has no effect at all,

(a) No magnetic field   (b) Instantaneous with magnetic field

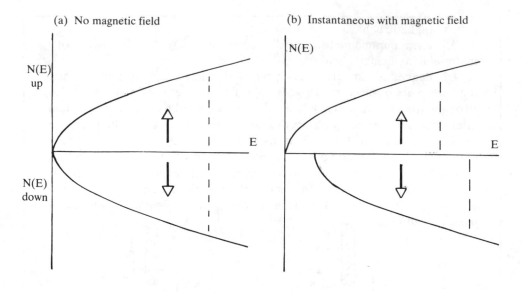

(c) Equilibrium with magnetic field

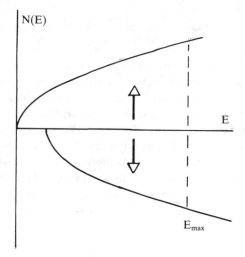

Fig. 3.11 — Metal electron energy levels, separating spin-up (top half) from spin-down electrons (bottom half). (a) With no magnetic field, (b) instantaneous effect of a magnetic field, (c) equilibrium effect of a magnetic field. The energy separation between the spin-up and spin-down electrons is much exaggerated. In the presence of an *electric* field, exactly the same separation and energy shift occurs between electrons moving to the *left* and to the *right*.

because there are no empty orbitals within that energy range to which the electron can be promoted. Thus, the only electrons which *can* be excited are those very close to E_{max}. Thermal excitation will therefore modify only the band occupancy from that shown in Fig. 3.10 to that in Fig. 3.12, in which the electrons in the shaded region

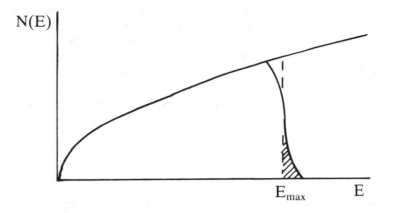

Fig. 3.12 — Effect of thermal excitation on electron energies near E_{max} (greatly exaggerated).

have been excited. The electronic contribution to the specific heat is so small because only these few electrons can take up thermal energy. Note the contrast between Fig. 3.12 and Fig. 3.11b, where a magnetic or electric field changes the energy of the whole body of electrons.

3.5 SUMMARY

This chapter has emphasized the strength of the simplest model of metallic bonding and the deductions that this leads to. Flourishing the alkali metals at every turn as examples has perhaps become a little monotonous! In reality, it is only for them that the model can provide accurate predictions. The fact that the five metallic electrons for vanadium lead to a calculated value of 18 eV for E_{max}, but the measured value is only 7 eV, shows the deficiencies of the simple approach. The excellent basis for understanding simple metals that the free electron model provides is modified in Chapter 6 to cover the more complex problem of metals with many valence electrons originating from d-orbitals.

3.6 APPENDIX: SOLUTION OF THE SCHRÖDINGER EQUATION FOR AN ELECTRON IN A BOX

3.6.1 One-dimensional box

The Schrödinger equation is a differential equation which governs the behaviour of electrons (Ladd 1979). In one dimension it has the form of (3.10)

$$(h^2/8\pi^2 m).d^2\psi dx^2 - (E-V)\psi = 0 \qquad (3.10)$$

(h = Planck's constant; m = mass of electron; E = total energy, V = potential energy). To apply the equation, the potential field experienced by the electron needs to be specified. One of the simplest potentials is that used in the free electron model of a metal (Fig. 3.2), with a uniform potential (which can be taken as zero) *within* the metal (between $x = 0$ and $x = a$) and an infinite potential outside it. To solve the Schrödinger equation, we need to find the functions $\psi_j(x)$ that satisfy the equation. Each of these describes one allowed wave-function (or *orbital*) for an electron. For each ψ_j, the corresponding E_j gives the energy of the electron in that orbital.

In general, to solve differential equations, a possible solution is guessed and its correctness tested. Here, a likely function will be one describing a wave, for example

$$\psi = A \sin(kx) + B \cos(kx) \tag{3.11}$$

where A, B and k are constants of unknown value. Differentiating this twice gives (3.12)

$$d^2\psi/dx^2 = -k^2 A \sin(kx) - k^2 B \cos(kx) \tag{3.12}$$
$$= -k^2 \psi$$

Substituting in (3.10) (with V taken as zero) gives

$$-h^2/(8\pi^2 m).k^2\psi - E\psi = 0 \tag{3.13}$$

that is, $E = k^2.h^2/(8\pi^2 m)$, the energy for a wave function of type (3.11) with a particular value of k (independent of A and B). If we consider the effect of the potential barriers, we see that not all values of k are possible. The electron cannot escape from the box. So,

$$\psi = 0 \text{ for } x < 0 \quad \text{and} \quad x > a.$$

More precisely,

$$\psi = 0 \quad \text{at} \quad x = 0 \quad \text{and} \quad x = a.$$

Substituting for x in (3.11) at $x = 0$, gives

$$\psi = 0 = A \sin(0) + B \cos(0).$$

As $\cos(0) = 1$, we deduce $B = 0$, that is, the wave function has only a sin term. Substituting at $x = a$, gives

$$\psi = 0 = A \sin(ka).$$

This is true only if $ka=n\pi$, where n is an integer (i.e. the function has a complete number of cycles within the box). Therefore all valid solutions are of the form

$$\psi_n = A \sin(n\pi x/a)$$

with

$$E_n = n^2 h_2/8ma^2. \tag{3.14}$$

Here n is the *quantum number* of the wave function. By normalization (Box 7), we find $A = \sqrt{2/a}$, that is

$$\psi_n = \sqrt{2/a}.\sin(n\pi x/a). \tag{3.15}$$

Fig. 3.13 shows the appearance of the three wave functions of lowest energy, those with $n = 1, 2, 3$.

3.6.2   Three-dimensional box
In a three-dimensional crystal, assumed for convenience to be a cube of side a, the Schrödinger equation becomes

$$(h^2/8\pi^2 m).(d^2\psi/dx^2 + d^2/dy^2 + d^2\psi/dz^2) - (E-V)\psi = 0 \tag{3.16}$$

Acceptable wavefunctions are specified by three quantum numbers, n_x, n_y, n_z and their energies are simply ((3.17), as (3.1))

$$E = (n_x^2 + n_y^2 + n_z^2)h^2/8ma^2 \tag{3.17}$$

3.6.3   Schrödinger equation for a hydrogen atom
The only difference between (3.16) for a cube and the corresponding equation for a hydrogen atom lies in the form of the potential energy, which is that due to the electrostatic attraction of the nucleus, that is,

$$V = -e^2/r.$$

The acceptable wave functions are much more complex that for an electron in a box, and are described in Box 6. The energy is comparatively simple

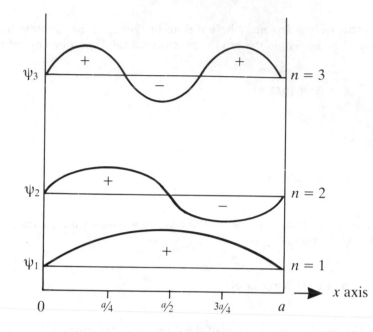

Fig. 3.13 — The first three wave functions for an electron in a uniform 1-dimensional potential:
$\psi_n = \sin(n\pi x/a)$ with $n = 1, 2, 3$.

$$E_n = 2\pi^2 e^4 m/n^2 h^2. \tag{3.18}$$

FURTHER READING

Metal structures

Adams, D. M. (1974) *Inorganic solids.* Wiley, Chichester.
Wells, A. F. (1984) *Structural inorganic chemistry.* Clarendon Press, Oxford (5th
    ed.)

Bonding and properties
Bube, R. H. (1981) *Electrons in solids.* Academic Press, New York, Chapter 6.
Coles, B. R. & Caplin, A. D. (1976) *Electronic structure of solids.* Edward Arnold,
    London, Chapter 3.
Cox, P. A. (1987) *The electronic structure and chemistry of solids,* Oxford Univer-
    sity Press, Oxford.
For a discussion of the mechanical properties of metals, see Guinier, A. & Jullien, R.
    (1989) *The solid state.* Oxford University Press, Oxford, Chapter 4

Text references
Ladd, M. C. F. (1979) *Structure and bonding in solid state chemistry.* Ellis Horwood,
    Chichester, p. 151f.
Pearson, W. B. (1985) *J. Appl. Cryst.* **18**, 540.

4

Ionic compounds

4.1 INTRODUCTION

Compounds in the top right corner of the bond triangle combine an electropositive, metallic element (A) with an electronegative non-metal (B) (Fig. 4.1). Because the B-atoms attract electrons strongly, the metallic bonding described in Chapter 3 is not possible. Instead, the A-atoms lose electrons, becoming A^+, and the B-atoms gain them, becoming B^- (or A^{2+}, B^{2-} if two electrons are lost or gained, etc.). For example:

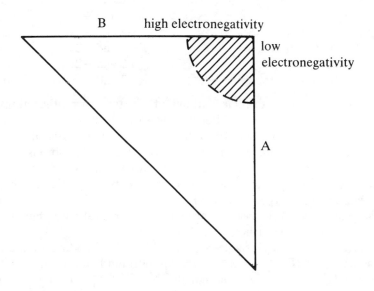

Fig. 4.1 — Ionic compounds in the bond triangle.

$$Na(metal) + 1/2Cl_2 \rightarrow Na^+ + Cl^- \rightarrow NaCl(solid)$$

These compounds are therefore made up of *ions*. They are held together by the electrostatic attraction between the positive and negative ions. They are called *ionic compounds* and their bonding is *ionic bonding*. As with metallic bonding, the electrostatic forces act in all directions rather than linking individual pairs of atoms. Calculation of the bonding energy of ionic compounds is relatively easy, and this leads to a good understanding of their chemical reactions, both at a qualitative and a quantitative level, in a way that is not possible for other classes of compounds. This aspect is therefore explored in some detail.

The classic examples of ionic compounds are the halides of the alkali and alkaline earth metals, such as NaCl and CaF_2. The oxides and fluorides of most other metals also behave as ionic compounds, as do a large group of mixed metal oxides (Chapter 10). However, compounds of metals (other than those in Groups I and II) with the heavier non-metals usually show large deviations from ionic character. This chapter assumes a simple description of ionic bonding involving complete transfer of electrons from A to B. Variations on this theme are examined in Chapters 10 and 11.

4.2 FORMULAE

One important property of ionic compounds is a result of their formation by transferring electrons from one atom to another. The ions have charges ±1,2,3, but the compounds themselves are neutral.† They must therefore be made up of precise numbers of cations and anions in the correct ratio. Consequently, ionic compounds of two elements can have only one of about five formulae:

AB when the charges are	+1 and −1	e.g. NaCl
or	+2 and −2	MgO
AB_2 when they are	+2 and −1	CaF_2
A_2B when they are	+1 and −2	Cu_2O
A_2B_3 when they are	+3 and −2	Al_2O_3
AB_3 when they are	+3 and −1	AlF_3

In metals it is easy to replace atoms of one type with other atoms, even if the numbers of valence electrons differ, because any extra electrons simply add to those delocalized through the solid. In contrast, in ionic compounds direct replacement of one ion, e.g. Na^+, by one of a different charge, e.g. Ca^{2+}, is impossible, even though they are the same size, because the extra positive charge cannot be neutralized. The same principles apply to compounds of more than two ions. Even though the range of formulae becomes large, they also have to contain precise ratios of the differently charged components. Table 4.1 shows some examples of mixed metal oxides, each of which has a distinctive structure.

Ions are not always single atoms. Thus, the bonding in metal nitrates or carbonates, e.g. $NaNO_3$, $CaCO_3$, and ammonium halides, NH_4X, is also of ionic character *between the ions,* though the NO_3^-, CO_3^{2-} and NH_4^+ ions themselves contain covalent bonds. These compounds also have simple formulae: AB, A_2B, etc.

---

† Higher charges are sometimes used for transition metal oxides (e.g. W^{+6} in WO_3), but these compounds are not really ionic.

Table 4.1 — Complex oxides

Name	Typical Formulae
Spinel	Fe_3O_4, $MgAl_2O_4$
Perovskite	$SrTiO_3$
Scheelite	$CaWO_4$
Zirconite	$ZrSiO_4$
Pyrochlore	$Ca_2Ta_2O_7$

for the same reasons. Even compounds in which the anions are complex polymers, including especially the silicates, have ionic bonding between the cations and anions, thought their properties are dominated by the character of the anions.

4.3 IONIC STRUCTURES

Ionic compounds take up structures that maximize their attractive energy, just as metals do. However, the ions experience both attraction and repulsion. We therefore expect them

(a) to maximize the attraction between positive and negative ions, but
(b) to minimize the repulsion between positive and positive or negative and negative ions.

Like metals, the ions approach each other as closely as possible until they reach the limiting distance where the electron clouds around each ion interfere and the repulsion is stronger than the attraction (Fig. 4.2).

The obvious way to achieve maximum attraction between equal-sized cations and

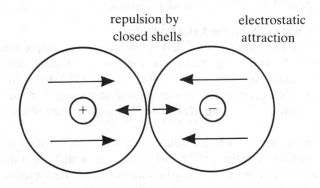

Fig. 4.2 — Limiting approach of cation and anion. The attractive force of the unlike charges (large arrows) is equal to the repulsive force due to interference between the electron clouds (small arrows).

anions would be to use the same packing as in metals, and place 12 anions around each cation. Fig. 4.3 shows one layer of this packing, with 6 anions around the cation;

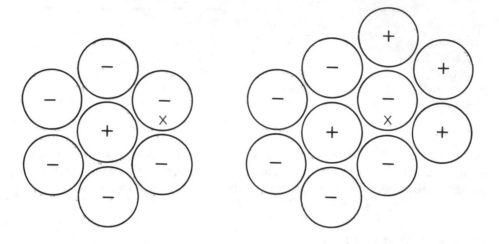

Fig. 4.3 — (a) Packing of a cation surrounded by 12 anions (6 in this layer, 3 each above and below). (b) Extension of the layer shown in (a).

3 more anions would be in the layer above and 3 below. However, this packing *does not work*. Ionic solids do not consist of isolated cations surrounded by anions. Consider what happens to anion (x) in Fig. 4.3a when the packing is extended (Fig. 4.3b). The best we can do in this layer is to give it 4 cation and 2 anion neighbours. This means that it is not possible to obtain satisfactory 12-coordinate environments for all the atoms in the structure. Solid ionic compounds must therefore have lower coordination than this. It is worth noting that if the ions are *isolated*, as in a coordination complex with anions around each cation, then 12-coordination is possible, e.g. in $[Ce^{3+}(NO_3^-)_6]^{3-}$, in which two of the oxygen atoms of each NO_3^- ion adjoin the Ce^{3+} ion.

4.3.1 [8:8] Coordination: the CsCl structure

The highest coordination number that is satisfactory in simple ionic compounds is eight. Fig. 4.4a shows how eight anions at the corners of a cube are packed around a central cation. Stacking eight of these cubes together (b) shows how each anion has eight cation neighbours in all, in an identical cubic arrangement to that of the anions around each cation. This eightfold coordination is described in brief as [8:8] coordination.

The structure just described is adopted by CsCl, CsBr, and CsI, and it is named from the first of these as the *CsCl* structure. Even though the ions have the highest possible coordination number, the structure is not at all common. The reasons for this are complicated and not fully understood. However, one important factor in controlling the choice of structure is certainly the size of the ions concerned. Consider what happens to the CsCl structure if the size of the cation is reduced. Starting from equal-sized ions (Fig. 4.5a), the cube shrinks until a critical size is

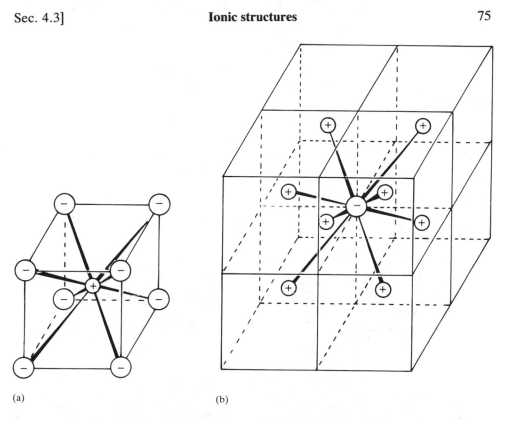

Fig. 4.4 — The CsCl structure (a) Eight-coordination of anions in a cube around a central cation. (b) The coordination of eight cations around each anion.

reached, when the ratio of cation radius to anion radius is 0.732:1 (Fig. 4.5b). At this ratio, the anions come into contact along the sides of the cube, and the cation just fits in the centre between them. Beyond this point the structure can be expected to become unstable; viewed naively, the cation 'rattles about' in a cavity that is too large for it.†

4.3.2 [6:6] Coordination: the NaCl structure
The preferred alternative to the [8:8] coordination of the CsCl structure has [6:6] coordination, in what is called the *NaCl* structure after its best-known example (Fig. 4.6a). Each cation (e.g. Na^+) has six anions (e.g. Cl^-) around it. These are at the corners of an octahedron (i.e. the positive and negative ends of the x, y, and z axes when the Na^+ ion is at the origin). Each anion is also surrounded by six cations in the same arrangement (Fig. 4.6b). This is extended in all directions to produce the full [6:6] NaCl structure (Fig. 4.6c).

---

† A more sophisticated investigation shows that for this particular structure, reduction in cation size does not produce the expected gain in binding energy. As a result, prediction of the expected structure of an AB compound from the radius ratio is not very successful. However, for other systems it can be a most effective guide to the structure (see, for example, Fig. 11.1).

(a)

(b)

(c)

(d)

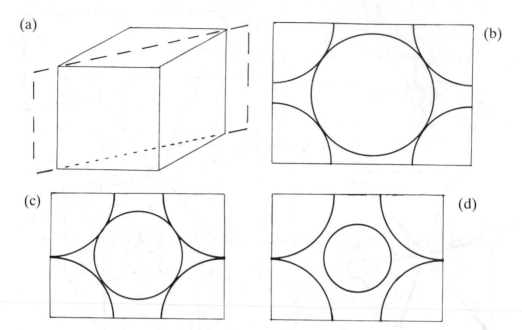

Fig. 4.5 — Diagonal plane through the CsCl structure (a) cubic unit cell showing location of the diagonal plane (b) Equal-sized ions (radius a_o); unit cell side 2.309 a_o (c) with cation:anion ratio of 0.732:1, and the anions just making contact; unit cell side 2.0a_o. (d) with cation:anion ratio of 0.6:1. Although the cations are smaller than in (b), the unit cell is still the same size.

As with CsCl, if the ratio of cation to anion radius in the NaCl structure falls below a critical value (0.414:1), the anions come into contact. Coordination [4:4] would then be preferred. Two very similar structures with this coordination do exist, known as *zincblende* and *wurtzite*, the two forms of ZnS (Fig. 4.7). However, most compounds with these structures have radius ratios well above 0.414:1. It is therefore clear that ion size is not the main factor controlling the stability of [4:4] in comparison to [6:6] structures. This problem is taken up in Chapter 10.

4.3.3 AB$_2$ structures
Ionic compounds of formula AB$_2$ show a similar range of structures. The highest coordination number is again eight, but now if eight B-anions coordinate each A-cation, there are only four cations for each anion. The coordination is therefore [8:4].† The type structure is that of fluorite (CaF$_2$) (Fig. 4.8a). This has an intriguing relationship to the [8:8] CsCl structure. The CsCl unit cube is doubled in each direction to give a block of 8 cubes. F replaces Cl at the corners of each cube, but Ca replaces Cs at the centres of *every other* cube, with the other cubes vacant. This gives the correct CaF$_2$ formula.

† Because each A-B interaction is also a B-A interaction, the ratio of coordination numbers (here 8:4) is always the inverse of the formula ratio (1:2).

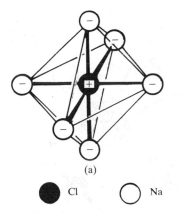

(a)

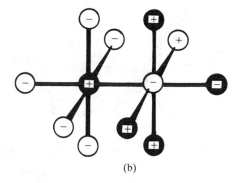

(b)

● Cl   ○ Na

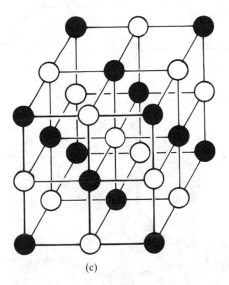

(c)

Fig. 4.6 — The NaCl structure (a) six anions (Cl⁻) around each cation (Na⁺) at the corners of an octahedron (b) six cations complete the coordination of each anion (c) the complete structure.

The F^- ions are of course still arranged in a cube around each Ca^{2+}, while the four Ca^{2+} around each F^- form a tetrahedron. This is shown in Fig. 4.8a(ii) for the central F^- ion; if the structure is extended in all directions, the same arrangement is found for all the other F^- ions.

The next step down from [8:4] coordination in AB_2 structures is to [6:3] coordination, and the same radius ratio rules can be applied to predict when it should occur. The standard structure for [6:3] coordination is that of rutile, TiO_2, which has octahedral and trigonal planar geometries for cation and anion respectively (Fig. 4.8b). Lower coordination still is found in some AB_2 compounds, for example [4:2] in

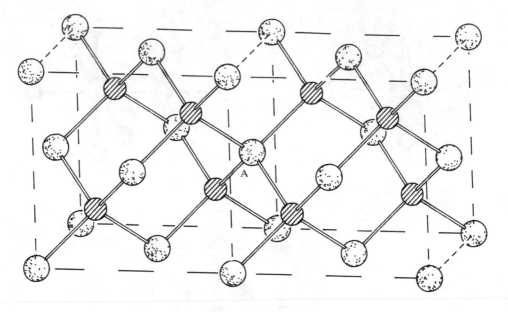

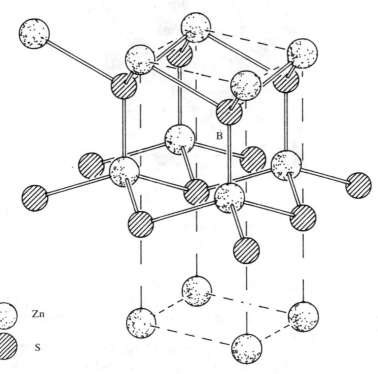

Fig. 4.7 — The structures of ZnS (a) the zincblende structure (cubic) (b) the wurtzite (hexagonal) structure. Both structures have the same [4:4] coordination, but differ in the arrangement of next nearest neighbours; In zincblende (a), linked tetrahedra are staggered (bond A); in wurtzite (b), they are eclipsed (bond B).

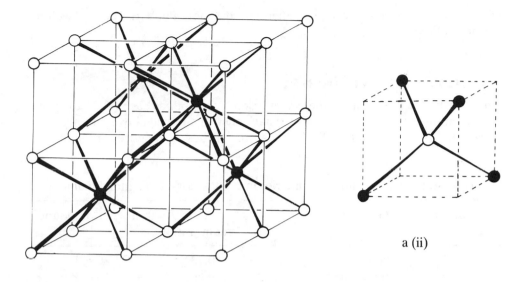

● Ca   ○ F

a (i)

a (ii)

b

● Ti   ○ O

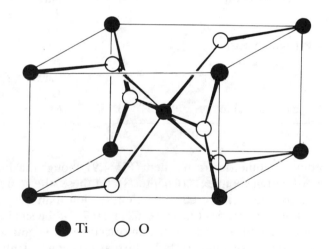

Fig. 4.8 — AB$_2$ structures (a) CaF$_2$ (i) unit cell made up of a block of eight cubes, with Ca at the centre of alternate cubes (ii) tetrahedral arrangement of four Ca^{2+} around the central F$^-$ (b) the TiO$_2$ (rutile) structure.

BeF$_2$. However, as with ZnS, the bonds in BeF$_2$ are probably not very ionic in character.

Ionic structures with other formulae, such as AB$_3$ or A$_2$B$_3$, can also be conveniently classified according to their coordination number. This is developed further in Wells (1984).

4.4 COORDINATION GEOMETRY

The coordination numbers in these simple ionic compounds each relate to specific coordination geometries. Most of these have been mentioned above, but it is useful to summarize them (Table 4.2). Two features of these geometries are surprising.

Table 4.2 — Coordination geometries in crystals containing individual ions

Coordination number	Geometry	Limiting radius ratio	Ion	Compound
8	cube	0.732	Cs,Cl	CsCl
8	dodecahedron	0.668	—	—
			F	CaF_2
8	square antiprism	0.645	Pb	PbClF
6	octahededral	0.414	Na,Cl	NaCl
			Ti	TiO_2
			Al	Al_2O_3
6	trigonal prism	0.528	Ni	NiAs
4	tetrahedron	0.225	F	CaF_2
			Zn,S	ZnS
			Be	BeF_2
3	trigonal plane	0.155	O	TiO_2

Three eight-coordinate geometries are frequent (Fig. 4.9), though the dodecahedron occurs only in coordination complexes (Chapter 8). Of these, the square antiprism and dodecahedron give much more effective coordination than the cube. This is shown by their limiting radius ratios (0.645, 0.668 and 0.732); the smaller this ratio, the smaller the cation can become before the anions come into contact. However, only the cube is at all common in ionic crystals. The reason is simple. Cubes can easily be attached to each other, but dodecahedra are almost impossible to link up into an extended crystal. Linking square antiprisms is difficult, because the widths at each end are different. The only ionic structures in which it occurs are of the PbClF type (including BiOCl, CaHCl, ThOS, and related compounds). The different-sized ions produce square ends of different widths, overcoming the square antiprism's packing problem.

    The two six-coordinate geometries have another cause. Octahedra, as in NaCl, TiO_2, etc. are much more numerous than the trigonal prisms of NiAs. This structure type is found for several transition metal sulphides, arsenides, etc. The most significant contrast between this and octahedral geometry is that the Ni atoms have other Ni atoms quite near them, through the triangular faces of the prism (shown

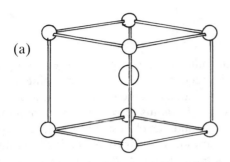

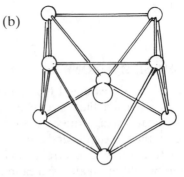

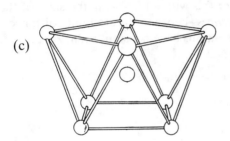

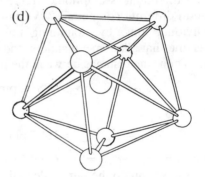

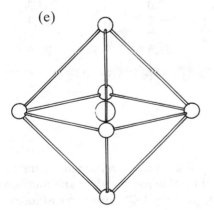

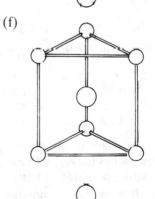

Fig. 4.9 — Eight and six-coordinate geometries (a) cube, e.g. CsCl (b) regular square antiprism, e.g. $[TaF_8]^{3-}$ (c) square antiprism with unequal ends, e.g. PbClF (d) dodecahedron, e.g. $[Mo(CN)_8]^{4-}$ (e) octahedron, e.g. NaCl (f) trigonal prism, as in NiAs, showing positions of adjacent Ni atoms.

in Fig. 4.9f). Thus, this structure is favoured when bonds can be formed between these atoms.

4.5   THE SIZE OF IONS

We have already assumed that the ions in ionic solids can be regarded as rigid spheres, each of a definite size. However, it is much more difficult to determine what these sizes are than is the case for metals. In a metal element, one atom touches another of the same type, and each can be assigned a radius corresponding to half the distance from centre to centre. In an ionic compound, the closest contacts are always between two different atoms: a cation and an anion. It is easy to determine the total distance, but much effort has gone into the attempt to divide this distance between the two ions, to obtain values of *ionic radius*.

Is the initial assumption correct? Are cations and anions reasonably constant in size in different compounds? We can answer this without knowing the sizes of the individual ions, by examining the *differences* between interatomic distances. Assume that alkali metal cations and halide anions have constant radii: r_{Na+}, r_{Cl-}, etc. Then, the distance between the ions in NaCl,

$$d_{NaCl} = r_{Na+} + r_{Cl-} = 282.0 \text{ pm (from X-ray studies)}$$

Similarly,

$$d_{KCl} = r_{K+} + r_{Cl-} = 314.7 \text{ pm}$$

Therefore $r_{K+} - r_{Na+} = 314.7 - 282.0 = 32.7$ pm

But we can do the same calculation for the other halides

d_{KF}	267.4	d_{KCl}	314.7	d_{KBr}	330.0	d_{KI}	353.3
d_{NaF}	231.0	d_{NaCl}	282.0	d_{NaBr}	298.7	d_{NaI}	323.7
	36.4		32.7		31.3		29.6

or, pairing the values differently, we can obtain $r_{Cl-} - r_{F-}$

d_{NaCl}	282.0	d_{KCl}	314.7
d_{NaF}	231.0	d_{KF}	267.4
	47.3		51.0

These important results show that ionic radii are indeed *reasonably* constant. From mean values for r_{M+} and r_{X-}, we should be able to reproduce the interionic distances with a precision of better than 3 pm, i.e. only 1–2% error in distances of 2–300 pm. It is, however, apparent that the remaining error is not random. The shorter interionic distances are slightly shortened, compared to the mean values, and the longer distances lengthened. Thus, the difference between KF and NaF is larger than that between KI and NaI. This is what would be expected if the ions are not precisely rigid, but are compressed by the ionic attraction, which is greater the smaller the ions involved. An interesting investigation has been carried out, to work out 'soft ionic radii' that allow for this (Holbrook *et al.* 1978). These radii do indeed predict interionic distances very well, but it is not clear how they can be used in interpreting structural trends, and they have not received the attention they perhaps deserve.

4.5.1 Ionic radii scales

When we know *one* ionic radius, say for Na^+, it is easy to obtain others, e.g. $r_{Cl-}=d_{NaCl}-r_{Na+}$; $r_{K+}=d_{KCl}-r_{Cl-}$, etc. A rather more sophisticated treatment avoids the arbitrary picking of NaCl+KCl rather than, say, NaF +KF to calculate r_{K+}. We can express all the radii in terms of one unknown value and minimize the errors in the predicted interatomic distances (Waddington 1966). The problem of finding this unknown value still remains. Various techniques have been used, notably by Pauling, by Goldschmidt, and by Landé, leading to the scales of ionic radius associated with each of their names. It is now generally agreed that all the earlier scales have been superseded by that based on direct measurement by X-ray methods. Precise calculations have been made from X-ray diffraction data (Chapter 3) of the electron density along the line joining cation to anion in various alkali halides. It is found that this density has a small minimum value, and it is reasonable to take this minimum as the transition point from cation to anion. Fig. 4.10 shows this minimum for LiF and compares it to the ionic radii used in various scales.

No published scale provides the ideal combination of the best base value obtained from X-ray study with the minimization of the deviations from additivity. However, an extensive compilation of radii has been made by Shannon (1976), using the interatomic distances in fluorides and oxides. This has been presented in two forms, based respectively on $r_{O^{2-}} = 140$ pm ($r_{F-} = 133$ pm), and on $r_{F-} = 119$ pm. The second set differ by only 1–2 pm from the ideal scale suggested above, and represent the best values currently available. Appendix A.4 provides a range of these values for 6-coordinate ions. It is finally worth noting that ionic radii, like metallic radii, are affected by coordination number, though reliable ratios have not been derived.†

4.6 PROPERTIES OF IONIC SOLIDS

The presence of ionic bonding leads to a number of properties characteristic to a greater or lesser extent of ionic solids. As well as having stoicheiometric formulae and distinctive structures, typically they are:

> Colourless
> Hard and brittle
> Unreactive
> High-melting.
> Insulating in the solid state, conducting when molten.

Each of these needs some qualification. The hardness and high melting points both reflect the strength of the ionic bonding. Thus, they are higher for small ions of high charge. Al_2O_3 has melting point 2045°C and hardness 9 on the Moh scale (running

---

† Tabulations of ionic radii in textbooks often use values from older compilations (as illustrated by Fig. 4.10), frequently without identifying their source . Published values should be carefully checked before they are used for calculation. In particular it is dangerous to combine values from different sources unless they are definitely consistent with each other.

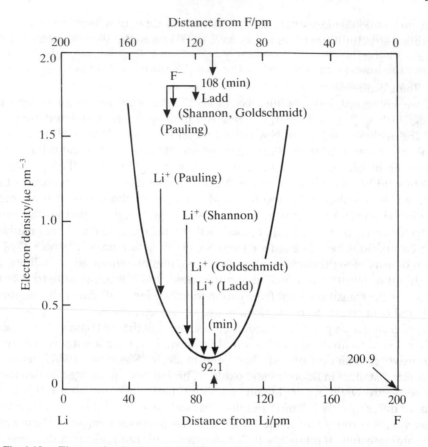

Fig. 4.10 — Electron density along the Li-F line in LiF, showing the positions of r_{Li+} and r_{F-} for various scales of ionic radius. Note that the maxima of electron density are far above the maximum values in the section drawn. Reproduced with permission from Greenwood, N. N., & Earnshaw, A. (1984) *Chemistry of the elements*, Pergamon Press, London, Fig.4.2.

from talc(1) to diamond(10)), while KCl with singly charged ions melts at 770°C and has a hardness of 2. The hardness relates principally to high shear strength. For two layers of an ionic crystal to slide over each other, the cation–anion bonds need to be broken completely before they can reform between different ions. This contrasts with metals where the bonding is little affected by the metal atoms moving relative to one another.

The property of brittleness causes materials to fail under tension far below their theoretical strength. These failures arise principally through cracks. The tip of a crack has a very small radius, and this leads to a concentration of stress, because the forces across the material have to pass around this tip. With ductile materials like metals, the forces reach the yield point of the material and flow occurs, increasing the radius of the tip of the crack and limiting the stress there. Non-ductile materials cannot flow readily, so cracks propagate when the stress at the tip of the crack exceeds the strength of the material (see Gordon 1979).

The lack of reactivity of ionic solids follows from their attainment of closed-shell electronic configurations, while the lack of colour indicates that the lowest empty electronic energy levels are considerably higher than the filled ones. In NaCl, for example, the lowest empty levels are the 3s levels of the Na atoms from which the electrons were transferred to the Cl atoms. Promotion to these levels requires UV radiation. Transition metal compounds deviate from both these properties because of the intrinsic character of the ions. They are frequently coloured because of their d–d or charge-transfer transitions (sections 7.3.1; 8.2.1) and in many cases undergo redox reactions. It is also noticeable that ionic compounds of post-transition metals are often coloured, e.g. yellow AgI. Such salts have smaller energy gaps between occupied and vacant orbitals. Absorption bands in the near UV extend into the blue, producing the yellow colour. In simple terms, the higher ionization potential of Ag (730 kJ mol^{-1} vs 490 kJ mol^{-1} for Na) stabilizes a promoted electron, reducing the energy gap.

The insulating or conducting properties of ionic solids are more complicated and are examined further in Part III. For NaCl and similar compounds, the insulating character arises because they lack delocalized electrons or any other charge carriers that can move through the crystal and conduct electricity.

4.7 THE BINDING ENERGY OF IONIC COMPOUNDS

An important feature of ionic compounds is the opportunity they offer to calculate their 'bond energy' from simple formulae not involving the quantum mechanics needed for covalent bonds. This leads to important insights into their chemistry. The bonds in ionic compounds are not directed from one atom to another, but act between all the atoms. The 'bond energy' is therefore the energy needed to break up the whole crystal into its component ions. This is known as the *lattice energy* (symbol U). It is defined as: the energy needed to convert one mole of crystalline solid to one mole of gaseous ions (in a thermodynamically standard state, i.e. non-interacting ions at standard temperature and pressure).† Numerical values of lattice energy are large and positive, e.g. +776 kJ mol^{-1} for NaCl.

The ease with which lattice energies can be calculated arises from a standard proof in the field of electrostatics. This states that the electrostatic effect of a charged sphere is exactly the same as the effect of the same charge concentrated at the centre of the sphere (Fig. 4.11).‡ Therefore, in dealing with a Na$^+$ ion, it can be treated simply as having a +1 charge located at its centre.

4.7.1 Electrostatic energy

The lattice energy of an ionic crystal is made up of various components, but the most important is that due to electrostatic forces. Taking NaCl as an example, the calculation starts by considering one pair of ions, Na$^+$ and Cl$^-$, at their standard distance apart in the crystal (r, 281.4 pm). The force between these ions is given by (4.1), the standard formula for electrostatic attraction between two point charges e at

---

† Some textbooks define lattice energy as the energy to *form* a crystal, i.e. giving it the opposite sign. The difference is unimportant, apart from the need to be consistent and unambiguous.
‡ This is proved by considering the charge distributed over the surface of the sphere and determining its effect at a point outside the sphere. Details are given in most physics textbooks.

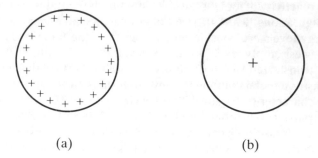

(a)   (b)

Fig. 4.11 — (a) Na^+ ion with positive charge distributed over the surface of a sphere (b) Equivalent model for calculation, with the charge concentrated at the centre of the sphere.

distance r. (ε_o is the 'permittivity of free space' which acts as a constant to give the result in SI units.)

$$F=-e^2/4\pi\varepsilon_o r^2 \tag{4.1}$$

If the ions are separated to an infinite distance, the energy needed is the work done against this force, i.e. the integration from r to ∞ (4.2)

$$U=\int Fdr=e^2/4\pi\varepsilon_o r \text{ joules} \tag{4.2}$$

We are interested in *one mole* of ions, so this is multiplied by N, Avogadro's number (4.3)

$$U=Ne^2/4\pi\varepsilon_o r \tag{4.3}$$

But, in NaCl (Fig. 4.7), we do not have simple pairs of ions, and we must consider the actual structure. Each Na^+ is surrounded by six Cl^- and so its electrostatic energy is six times that given by (4.3) (Fig. 4.12a). However, it also interacts with the twelve

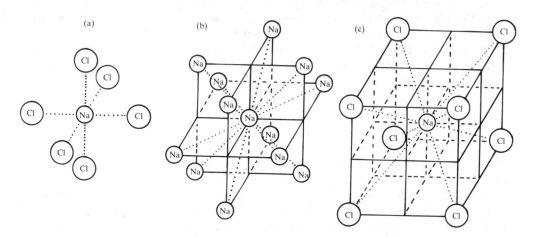

Fig. 4.12 — Interactions in NaCl (a) attraction to six Cl^- distance r (b) repulsion by twelve Na^+ distance $\sqrt{2}r$ (c) attraction by eight Cl^- distance $\sqrt{3}r$.

Na$^+$ ions that are rather further away ($\sqrt{2}r$) and these, of course, repel the original Na$^+$ ion, *reducing* the overall energy by the amount given by (4.4) (Fig. 4.12b).

$$12\ Ne^2/4\pi\varepsilon_0\sqrt{2}r \tag{4.4}$$

Further away still are eight Cl$^-$ at the corners of each of the small cubes around the Na$^+$, distance $\sqrt{3}r$ (the length of the body diagonal of the cube of side r). These add to the attractive part of the energy. It is worth remembering that we are considering point charges. Therefore no shielding effects arise and all interactions need to be considered, even between two Na$^+$ with a Cl$^-$ directly between them.

The energy of the whole crystal is the sum of these various terms, considering all the interactions out to an infinite distance, i.e. (4.5) (separating out the common factors)

$$U=NE^2/4\pi\varepsilon_0 r\{6-12/\sqrt{2}+8/\sqrt{3}-6/\sqrt{4}+\ldots\} \tag{4.5}$$

The bracketed series is difficult to sum because it converges slowly, but its value was obtained by the German physicist E. Madelung as 1.7476... This sum is therefore called the *Madelung constant* (M).

It is important to note that the Madelung constant depends on the structure. Thus, for CsCl (Fig. 4.13)

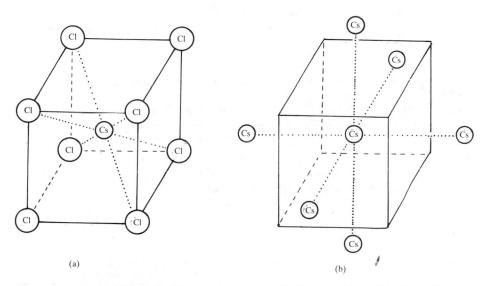

Fig. 4.13—Interactions in CsCl (a) attraction to eight Cl$^-$ distance r (b) repulsion by six Cs$^+$ at the centres of each adjacent cube, distance $2r/\sqrt{3}$.

$$M=8-6*2/\sqrt{3}+\ldots=1.763. \tag{4.6}\dagger$$

This series was also summed by Madelung. For more complex structures, particu-

† The difference between M for NaCl and CsCl is only 1%. When the increased radii of 8-coordinate rather than 6-coordinate ions are taken into account, the CsCl structure is hardly if at all more stable than NaCl for a given pair of ions. This is one of the factors underlying the general preference of AB compounds for the NaCl rather than the CsCl structure.

larly if the unit cell is not cubic, the sum can be obtained only by computer (and then only with difficulty because of the slow convergence). Computer summation has even been applied to compounds of complex ions such as $CaCO_3$, though this is a difficult calculation requiring the distribution of the 2- charge of the CO_3^{2-} over the four atoms of the anion.

The energy calculation needs to be modified if the ions do not have charges +1 and -1. MgO has the NaCl structure, with $r = 210.6$ pm. The force between its ions is

$$2*2*e^2/4\pi\varepsilon_o r^2 \tag{4.7a}$$

and the electrostatic energy is

$$U = N*1.748*4*e^2/4\pi\varepsilon_o*210.6 \tag{4.7b}$$

In general the formula for the electrostatic energy when the ions have charges z_+ and z_- is (4.8)

$$U = NMz_+z_-e^2/4\pi\varepsilon_o r \tag{4.8}$$

To give an indication of the magnitude of U, inserting values for the various constants for an NaCl structure gives (with +1 and -1 charged ions)

$$U = 245\ 100/r \text{ kJ mol}^{-1} \ (r \text{ in pm})$$

i.e. $U = 869$ kJ mol^{-1} for NaCl and
            $U = 4655$ kJ mol^{-1} for MgO ($z_+ = z_- = 2$).

Table 4.3 summarizes values of the Madelung constant for different ionic

Table 4.3 — Madelung constants for ionic structures

Formula	Name	Madelung constant
CsCl		1.763
NaCl		1.748
ZnS	Zincblende	1.638
	Wurzite	1.641
CaF_2	Fluorite	2.519
TiO_2	Rutile	2.408[a]
Al_2O_3	Corundum	4.172[a]

[a] These values vary slightly for different compounds, depending on the precise shape of the unit cell (the ratios between the axis lengths). The quoted values relate to TiO_2 and Al_2O_3 themselves.

structures.† The larger values for the more complex structures arise because one mole of, say, Al_2O_3 contains 5 ions compared to 2 for NaCl. Irrespective of the structure, the Madelung constant *per ion* is 0.8–0.85. This represents the increase in

† Other values are frequently quoted for these constants, usually because the charges z_+ and z_- have been incorporated into the Madelung constants. This is confusing, but not incorrect as long as (4.8) is modified appropriately. However, in many texts, the form of (4.8) used is stated for only the +1, -1 case, or is actually inconsistent with the values of M quoted. See Quane, (1970).

electrostatic energy on forming a crystal, compared to the value of 0.5 per ion (1.0 for an ion pair) for isolated ion pairs. The constancy of the Madelung constant has been used by Kapustinskii in a general equation (4.9) giving an approximate lattice energy for any ionic solid, irrespective of its structure:

$$U = 120\ 050\ Vz_+z_-/(r_++r_-)(1-34.5/(r_++r_-)) \tag{4.9}$$

where V is the number of ions per mole (e.g. 2 for NaCl, 3 for CaF_2), and r_+ and r_- are the cation and anion radii; the term $(1-34.5/(r_++r_-))$ corrects for the repulsive energy, as discussed next.

4.7.2 Repulsive energy

If only electrostatic forces were present in a crystal, it would contract to an infinitesimal point, to give infinite electrostatic energy. As this does not happen, we can deduce that at the equilibrium distance between the ions (r_0), the electrostatic forces are exactly balanced by repulsive forces due to the interaction of the electron clouds on each ion. Because these are short-range, their effect on the overall energy is quite small, i.e. only a small amount of work is done by these forces when the crystal expands into a gas. This can most easily be calculated as a correction to U, assuming the expression used by Born and Mayer for the *repulsive energy* RE (4.10)

$$RE = -a.\exp(-r/\rho) \tag{4.10}$$

where a and ρ are constants. Then

$$U = NMz_+z_-e^2/4\pi\varepsilon_o r - a.\exp(-r/\rho) = A(\text{electrostatic}) - B(\text{repulsive}) \tag{4.11}$$

We know that at equilibrium (when $r=r_e$), repulsive and electrostatic *forces* are equal. Remembering that

$$E = \int F dr \text{ or } F = dE/dr,$$

the first derivatives of A and B are equal and opposite, i.e.

$$NMz_+z_-e^2/4\pi\varepsilon_o r_o^2 = -1/\rho.a.\exp(-r/\rho) \tag{4.12}$$

or

$$-A/r_o = -B/\rho$$

$$B = -\rho/r_o.A$$

Thus, $U\ (=A-B)$ is given by (4.13)

$$U = NMz_+z_-e^2/4\pi\varepsilon_o r_o(1-\rho/r_o) . \tag{4.13}$$

Note that this formula is correct only at the equilibrium distance, r_o, though the electrostatic term is valid at any distance. The constant ρ has been estimated from quantum-mechanical calculations as 34.5 for all alkali metal halides.†

For precise lattice energy calculations, two further energy terms need to be included. All atoms, whether charged or not, attract each other weakly, by *van der*

---

† An alternative equation for the repulsive energy, $RE = -c/r^n$, leads to a correction term $(1-1/n)$. However, n varies from 5 to 10 for the alkali metal halides, making this a less well-defined correction.

Waals forces (discussed in section 7.2). The size of the corresponding energy term (usually called the *dispersion* or the *London energy*) depends mainly on the polarizability of the ions, but it can be comparable to the repulsive energy for salts of large easily polarizable ions (e.g. CsI). Finally, all solids have *zero-point energy,* the residual vibrational energy of the lattice at 0 K. This small term can be calculated from the vibration frequencies of the lattice. Typical values for these energies are given in section 4.8.1. Dasent (1970, pp. 76–8) gives details of the calculation of these minor terms.

4.8   APPLICATIONS OF LATTICE ENERGY CALCULATIONS

The formula for lattice energy is used in two ways. Firstly, the actual numerical values of U for particular compounds can be calculated or applied. Secondly, the relationship $U \sim 1/r$ can explain trends in chemical behaviour related to the sizes of the ions.

4.8.1   Comparison of experimental and calculated U
An important and obvious numerical application is to compare the calculated U with the value that can be obtained experimentally by thermochemistry. The lattice energy cannot be measured directly, but it can be obtained by using Hess's Law (that the heat change in a reaction is unaffected by the route taken from reactants to products). In the thermochemical cycle shown in Fig. 4.14 (the Born-Haber cycle),

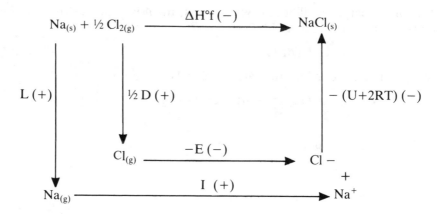

Fig. 4.14 — Born-Haber cycle for the formation of solid NaCl. Each term has (+) or (−) after it, to indicate if the energy change is endo- or exothermic in the direction shown. U=lattice energy; L=latent heat of sublimation of Na; D=dissociation energy of Cl_2; E=electron affinity of Cl; I=first ionisation potential of Na.
Note that this is an enthalpy (ΔH) cycle, whereas U measures a free energy (ΔG) change. This requires the correction

$$\Delta H = \Delta U + P\Delta V$$

i.e. 'lattice enthalpy'='lattice energy'(U)+nRT for the expansion of n moles of solid ions to 1 atm. pressure ($RT \sim 2.5$ kJ mol^{-1}).

the value of U is the only unknown quantity. The heat of formation of solid NaCl is given by (4.14)

$$\Delta H_f^\circ = L + 1/2D + I - E - (U + 2RT) \tag{4.14}$$

and the experimental value of U can therefore be calculated from (4.15)

$$U_{BH} = I + L + 1/2D - E - \Delta H_f^\circ - 2RT \tag{4.15}$$

For NaCl, the numerical values are

$$U_{BH} = 502 + 108 + 121 - 354 + 411 - 5 = +782 \text{ kJ mol}^{-1}.$$

The calculated value of U is made up of the following energies:

Electrostatic	863
Repulsion	-114
Dispersion	25
Zero-point	8
Total	776 kJ mol^{-1}.

The agreement is excellent, well within the experimental error for the two methods. It is important to realize that this does not *prove* that the ionic model is correct. This can be done only by using X-ray methods to show that the electron distribution corresponds to that assumed by the model (as indeed seems to be the case). However, the agreement in energy does give confidence in the calculated values of U. For appropriate compounds it should be possible to calculate accurate 'bond energies' when experimental values are unobtainable.

Table 4.4 shows the same comparison for some other compounds. For CsI, the

Table 4.4 — Comparison of Born-Haber cycle and calculated lattice energy values (kJ mol $^{-1}$)

Compound	Born–Haber cycle	Calculated	Difference
NaCl	782	776	6
CsI	602	592	10
AgF	953	920	33
AgCl	903	832	69
AgBr	895	815	80
AgI	882	777	105
CuCl	978	903	75
CuBr	907	869	99
CuI	957	832	125

figures are still very close. However, the Ag^+ and Cu^+ halides tell a different story. Obviously, their bond energies are not completely explained by the ionic model. In particular, it is significant that as the ion sizes increase, their lattice energies hardly decrease at all. This shows that some covalent bonding must be present, involving the interaction of electrons on X^- with vacant orbitals on Ag^+ or Cu^+. This covalent contribution cannot easily be determined because the covalent bonding reduces the effective charges on each ion, and thus the ionic lattice energy. Therefore, the covalency must provide substantially more energy than the differences shown in Table 4.4. Covalent bonding in predominantly ionic compounds is discussed further in Chapter 11.

Ag^+ and Na^+ are almost the same size. Their salts would therefore be expected to have similar solubilities. In fact, AgCl, AgBr, and AgI are notoriously insoluble. If the thermodynamics of the process are examined, it is found that the heats of solution of the Ag^+ salts are much lower than for the corresponding Na^+ salts. The differences between these values are almost identical to the differences in observed and calculated lattice energies for the Ag^+ salts, i.e. the insolubility is caused by the same partial covalent bonding that causes the lattice energy discrepancies.

4.8.2 Determination of terms in the Born–Haber cycle

Calculated U values are essential when one of the other terms in the Born–Haber cycle is uncertain. For some time, the values obtained by spectroscopic methods for the electron affinity of fluorine

$$F + e^- \rightarrow F^-$$

were suspect, until they were confirmed by using calculated U values. Similarly, the two-electron affinity of oxygen

$$O + 2e^- \rightarrow O^{2-}$$

is endothermic ($-670\,kJ\,mol^{-1}$). It cannot therefore be obtained experimentally and has to be calculated from a Born–Haber cycle.

A similar technique is applied to find the *proton affinity* (P) of NH_3

$$NH_3 + H^+ \rightarrow NH_4^+.$$

The NH_4^+ ion can be treated as a spherical atom of similar size to K^+. The necessary cycle (Fig. 4.15) gives

$$P = -U - 2RT + I - E + \Delta H_f^\circ(NH_3) + 1/2D(H_2) + 1/2D(X_2) - \Delta H_f^\circ(NH_4X),$$

and substitution of numerical values gives the results in Table 4.5

These show excellent consistency for NH_4Cl, NH_4Br, and NH_4I, indicating a value of $876\pm2\,kJ\,mol^{-1}$ for the proton affinity. The value for NH_4F is about 75 kJ mol^{-1} bigger, a discrepancy that arises because the lattice energy calculation has omitted one factor. Ammonium fluoride contains strong hydrogen bonds (section 7.4), N–H . . . F. These cause the structure to adopt the ZnS (wurtzite) structure (Fig. 4.7b), in which each tetrahedral NH_4^+ ion is surrounded by four F^- ions, perfectly placed to form these hydrogen bonds. In the other ammonium halides at room temperature, the NH_4^+ ions rotate freely, behaving as if they were spheres. These

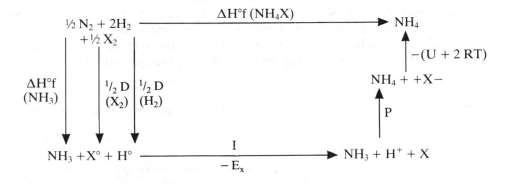

Fig. 4.15 — Born-Haber cycle for the formation of NH_4X (X=hal). P=proton affinity of NH_3, defined as for electron affinity, i.e. positive for an exothermic change.

Table 4.5 — Calculation of proton affinity of NH_3 (kJ mol^{-1})

Salt	Proton affinity	Structure	Coordination numbers
NH_4F	951	wurtzite	4:4
NH_4Cl	875	NaCl	6:6
NH_4Br	880	NaCl	6:6
NH_4I	872	NaCl	6:6

salts therefore have NaCl structures (though at low temperature the rotation ceases and they transform to the wurtzite type).

The energy of the N–H . . . F bonds which was not included in the lattice energy calculation is about 75 kJ mol^{-1}. Thus, each of the four individual hydrogen bonds has a bond energy of about 18 kJ mol^{-1}. The determination of this value (otherwise difficult to obtain) is an unexpected bonus from the proton affinity calculation. Similar calculations give the fluoride ion affinity of BF_3, the enthalpy of the reaction

$$BF_3 + F^- = BF_4^-.$$

4.8.3 Unknown compounds

Another quantitative application of lattice energy calculations is to the estimation of the heat of formation of an unknown compound. If this is assumed to be ionic, its lattice energy can be calculated and the heat of formation obtained from a Born–Haber cycle. Good examples of this are the Cu, Ag, and Au monofluorides. Although all the other halides of these metals are known, the only fluoride is AgF.

The radii of Cu^+ and Au^+ can be estimated from these other halides, and, like AgF, they would probably have NaCl structures. The calculated lattice energies should be reasonably accurate because the discrepancies for the other Cu and Ag halides do not affect AgF substantially. It is instructive to compare the Born–Haber calculations for AuF with that for KF, which has a cation of identical size (equation (4.14) and Table 4.6).

Table 4.6 — Comparison of Born–Haber cycles for AuF and KF (energies in KJ mol^{-1})

Compound	$U+2RT$	L	$1/2D$	I	E	ΔH_f°
AuF	777	368	77	894	−339	+223
KF	801	90	77	419	−339	−566

It is obvious that AuF does not exist, because it is thermodynamically unstable to decomposition to the elements (positive ΔH_f°). The reason for this is also clear. Both I and L are much larger for Au than for K, reflecting that Au is less electropositive than K, and that it has much stronger metal–metal bonding. In contrast, both CuF and AgF are exothermic (-59 and -205 kJ mol^{-1}), principally because the smaller cation sizes produce larger lattice energies which are high enough to compensate for I and L.

The instability of CuF must therefore have a different cause. Consider the disproportionation of CuF

$$2CuF = Cu + CuF_2$$

Knowing the standard heat of formation of CuF_2 (530 kJ mol^{-1}), we can calculate the heat change for this reaction as

$$\Delta H = \Delta H_f^\circ(CuF_2) - 2\Delta H_f^\circ(CuF) = -530 - 2*(-59) = -412 \text{ kJ mol}^{-1}$$

The disproportionation is strongly favoured, pinpointing the stability of Cu^{2+} as the Achilles heel of CuF. In contrast, though AgF_2 exists, it is thermodynamically fairly unstable.

4.9 QUALITATIVE USE OF LATTICE ENERGY CALCULATIONS

In its simplest form, the lattice energy of a compound AB_n can be approximated by (4.16)

$$U_{AB} = k/r_{AB} \tag{4.16}$$

where r_{AB} is the distance between the ions (Fig. 4.16a). This can be used to understand reactions in which one solid is converted into another.

4.9.1 Decomposition of complex ions
An important class of solid state reactions involves the decomposition of a complex ion, e.g. in calcium carbonate

$$CaCO_3 \rightarrow CaO + CO_2$$

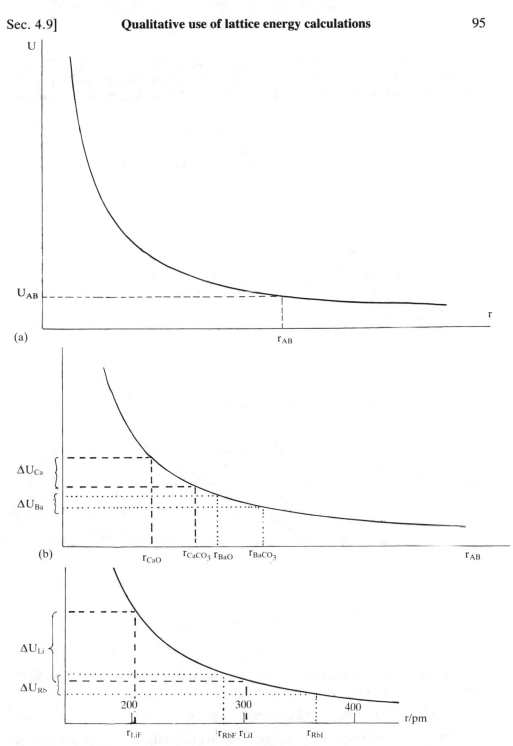

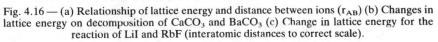

Fig. 4.16 — (a) Relationship of lattice energy and distance between ions (r_{AB}) (b) Changes in lattice energy on decomposition of $CaCO_3$ and $BaCO_3$ (c) Change in lattice energy for the reaction of LiI and RbF (interatomic distances to correct scale).

It is useful to discover whether the carbonates of larger ions, e.g. $SrCO_3$ or $BaCO_3$, will be more or less stable than $CaCO_3$. The decomposition reaction can be included in a thermochemical cycle (Fig. 4.17a). In this cycle, ΔH_g is the *gas-phase* decompo-

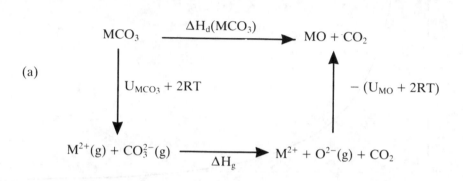

(a)

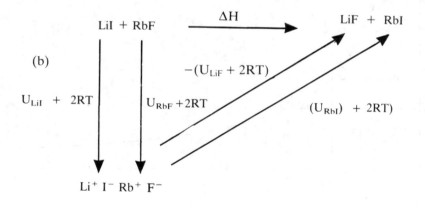

(b)

Fig. 4.17 — Thermochemical cycles for solid-state reactions, (a) decomposition of a carbonate, (b) double decomposition of LiI and CsCl.

sition enthalpy of the carbonate ion, which is independent of the metal involved. Then, the enthalpy of the reaction is given by

$$\Delta H_d(MCO_3) = U_{MCO3} - U_{MO} + \Delta H_g. \tag{4.17}$$

Thus, the ease of decomposition depends only on the difference in lattice energy between the carbonate and the oxide. We can assume that decomposition of the carbonate causes a uniform reduction in the interionic distance, i.e.

$r_{MCO3} - r_O =$ constant (d)

Fig. 4.16b compares the effect of this decomposition on the carbonate of (a) a small cation, e.g. Ca^{2+} and (b) a large cation, e.g. Ba^{2+}. It is clear that for the small cation, the change in distance (d) produces a much greater change in U. From equation (4.17), we see that $BaCO_3$ should be more stable than $CaCO_3$. This is completely confirmed by the decomposition temperatures:

Stability	$BaCO_3$	$>SrCO_3$	$>CaCO_3$	$>MgCO_3$	$>BeCO_3$
Decomposition temp(°C)	1360	1189	897	540	<20

At the decomposition temperature, ΔG $(=\Delta H - T\Delta S)$ becomes negative. The carbonate ion has a positive ΔS for decomposition, because a molecule of gaseous CO_2 is produced. Therefore, the more stable the carbonate, the higher the temperature has to be before ΔG becomes negative.

This behaviour can be expressed in a general rule: *A complex cation or anion is stabilized by combination with a large anion or cation* (respectively). Examples are very numerous, and include the stabilization of ICl_4^- and HF_2^- with K^+ or larger cations, and the formation of PH_4^+ as the bromide or iodide but not the chloride. The alkali metal nitrates show a similar sequence to the carbonates, but with two decomposition steps

$$MNO_3 \rightarrow MNO_2 \rightarrow M_2O.$$

Burning the alkali metals in air produces superoxides, peroxides, or oxides:

CsO_2, RbO_2, KO_2; Na_2O_2; Li_2O
superoxides   peroxide   oxide

This again relates to the lattice energy stabilization of the complex ions, though the cycle for the decomposition of a superoxide is slightly more complex than that in Fig. 4.16b, because two moles of superoxide produce one mole of peroxide

$$2KO_2 \rightarrow Na_2O_2 + O_2$$

4.9.2 Double decomposition of ionic solids
The simplest possible ionic reaction is that between two ionic compounds in the solid state, possibly leading to the interchange of the ions

$$AB + CD \rightarrow AD + CB$$

Such reactions take place quite readily. For example, LiI and RbF can be ground together and heated, and their X-ray powder pattern observed. Quite rapidly, the lines due to these two compounds will decrease in intensity, and new lines appear, recognizable as those of LiF and RbI. Tabulated heats of formation for the four compounds shows that the reaction

$$LiI + RbF \rightarrow LiF + RbI$$

is exothermic ($\Delta H = -18$ kJ mol^{-1}), and this can be understood from the thermochemical cycle in Fig. 4.17b. The heat of reaction depends only on the lattice energies

$$\Delta H = -U_{AD} - U_{CB} + U_{AB} + U_{CD}.$$

The relationship of this to the size of the ions can be followed in Fig. 4.16c. It is obvious that the combination LiF+RbI is favoured by the trend in lattice energy. The graph shows that the cation/anion combinations of small+small/large+large will always be favoured over the other pairing.†

Both this behaviour and the stability of complex ions are sometimes explained in terms of 'two large ions fitting together better than a large ion with a small ion'. It is important to realise that not merely is this nonsense, but that it is unnecessary. The chemical behaviour is completely explained by the effects of changing ion size on lattice energy.

4.9.3 Solubility of ionic compounds

The relationship between lattice energy and ion size can also be used to investigate the trends in solubility in ionic salts, as the cation or anion size changes. The solubility depends on the balance of three factors: the lattice energy, and the solvation energy of the cation and of the anion. While the lattice energy is related to the sum of the radii ($k/(r_+ + r_-)$; 5.16), the solvation energies depend on the individual radii ($p/r_+ + p/r_-$), where k and p are constants (Morris 1968). Differentiation of ($k/(r_+ + r_-) - p/r_+ - p/r_-$) with respect to r_+ or r_- leads to the prediction that the solubility should have a parabolic form, first decreasing then increasing as the cation size increases. The experimental results fit well with this prediction, though for most anions the range of cation size is too small to show the complete curve. Thus, for Br^-, I^- and other large anions the solubility order of the alkali metals is Li>Na>K ..., i.e. one side of the parabola; for F^- and OH^-, the order Li<Na<K ... shows the opposite trend, while for Cl^-, the solubility has a minimum. For a detailed study of solubility, the reader is referred to Johnson (1982).

FURTHER READING

Structures

Adams, D. M. (1974) *Inorganic solids*. Wiley, London, Chapter 3

Wells, A. F. (1984) *Structural inorganic chemistry*. Clarendon Press, Oxford, Chapter 6.

Ionic Radii

Waddington, T. C. (1966) *Trans. Faraday Soc.* **62** 1482.

Shannon, R. D. (1976) *Acta Cryst.* **A32**, 751.

Properties

Gordon, J. E. (1979) *The new science of strong materials*. Pitman, London. Most inorganic chemistry texts discuss the general properties of ionic compounds. This ref. gives a detailed examination of the strength of solids.

---

† Unless the lattice energy of one of the compounds contains a non-ionic contribution. AgI+CsF is more stable than AgF+CsI, despite the size of the ions, because of the non-ionic component in the lattice energy of AgI.

Energetics
Burgess, J. (1978) *Metal ions in solution.* Ellis Horwood, Chichester, Chapter 7.
Dasent, W. E. (1970) *Inorganic energetics.* Penguin, Harmondsworth.
Johnson, D. A. (1982) *Some thermodynamic aspects of inorganic chemistry.*
Cambridge University Press, Cambridge.
Morris, D. C. F. (1968) *Ionic radii and enthalpies of hydration of ions. Structure and bonding.* **4** 63.

Text References
Holbrook, J. B., Khaled, F. M. & Smith, B. C. (1978) *J. Chem. Soc. Dalton Trans.*
p. 1631.

5

Covalent compounds

5.1 INTRODUCTION

The last two chapters have examined compounds formed either by two electro-positive atoms, giving metallic bonding, or with one electropositive and one electronegative atom, giving ionic bonding. When neither element is electropositive, the compounds are found in the lower right corner of the bond triangle (Fig. 5.1).

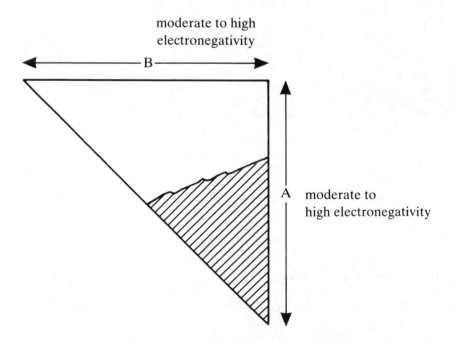

Fig. 5.1 — The bond triangle, showing the area in which covalent compounds are found.

Rather by default, they form covalent bonds, because both atoms hold electrons too strongly either to lose them in metallic bonding, or to transfer them from one atom to another in ionic bonding. Covalent bonding can exist even if the elements involved differ considerably in their electronegativity, so it covers a large part of the bond triangle. Indeed, as will be seen in Part III (Chapters 7 and 12), it sometimes seems that metallic bonding and even ionic bonding should be regarded as special types of covalency. Here we are concerned with much more straightforward examples: two rather electronegative atoms (i.e. *non-metals*) in combination.

We take, as a starting point, the simplest pair: two atoms of hydrogen, each with one electron in its 1s orbital (Box 6). What happens as they approach each other?

This question has been answered in many ways — from the simplest 'hook-and-eye' bonding picture, to the precise calculation, using rigorous quantum mechanical formulae (and many hours of computer time) of the energy and electron distribution in the combination of two protons and two electrons. We will describe the bonding in terms of *molecular orbitals* — the orbitals occupied by the electrons in the *molecule*, usually with the atoms in their final bonded position.

5.2   CONSTRUCTION OF MOLECULAR ORBITALS

We need not guess at the nature of these orbitals, but will *construct* them from the interaction of the original atomic orbitals as the atoms approach each other. It has been found that the molecular orbitals can be formed by adding or subtracting the atomic orbitals, i.e. forming *linear* combinations (rather than, say, quadratic combinations involving the squares of the original orbitals). The procedure is therefore called the:

> *Linear combination of atomic orbitals* (LCAO) method.

The molecular orbitals produced by this method obey various rules. The most important is:

1. The number of molecular orbitals equals the number of contributing atomic orbitals.

The second rule is stated for two atomic orbitals, but applies, more broadly, to combinations of several orbitals.

2. If two atomic orbitals interact at all (section 5.5.1), one of the two molecular orbitals formed will be of higher energy, and one of lower energy.

To a first approximation, the average energy of the molecular orbitals equals the average energy of the atomic orbitals. More detailed analysis shows that the upper level rises rather more than the lower one drops (Box 7).

5.3   MOLECULAR ORBITALS FOR H$_2$

How does the LCAO method apply to H$_2$ and how can the molecular orbitals be described? They are represented in two entirely different ways.

Box 6 Atomic orbitals

The solutions to the Schrödinger equation for the hydrogen atom (section 3.6.3) define the allowed wave functions for its single electron. These wave functions are the *atomic orbitals*. They are identified by three quantum numbers:

n principal quantum number $n=0, 1, 2, ..$
l azimuthal quantum number $l=0, 1, 2, .., n-1$.

The numbers are generally replaced by letters: $0=, 1=p, 2=d, 3=f$.

m magnetic quantum number $m=0, \pm1, .. \pm1$.

This means that one s, three p, five d, and seven f orbitals exist. Additionally, every electron has a spin quantum number, $s=\frac{1}{2}$ or $-\frac{1}{2}$. By the *Pauli principle*, all electrons in an atom must have different quantum numbers. Thus each orbital (defined by a set of n, l, m values) can contain two electrons, one with $s=\frac{1}{2}$ and one with $s=-\frac{1}{2}$.

Orbitals are generally labelled with the Greek symbol ψ. This stands for the mathematical function describing the movement of the electron around the nucleus. For the hydrogen atom, we known the exact form of these functions. For example, the H-1s orbital depends only on the distance (r) from the nucleus

$$\psi(r)=2a_0-6.5\exp(-r/a_0)$$

where a_0 is a constant. For other atoms, we generally assume the orbitals are the same *shapes* as for hydrogen, although their energies are different; the values of the functions also be calculated point-by-point.

For the study of bonding, the shapes of the outermost, part-filled atomic orbitals are most important. These are the *valence orbitals*. They can be represented in many ways. Figure B6 shows the variation of ψ with angle.

Three features of these diagrams are particularly important. Each lobe has a *sign*, which represents the phase of the wave function in that part of the orbital. When orbitals overlap, lobes with the same sign reinforce each other, while those with different signs cancel each other out.

The surface where an orbital changes sign is a *node*, often a *nodal plane*, but sometimes a more complex surface.

Finally, the orbitals can be divided into two groups: *centrosymmetric*, which have identical values at points related by the centre of symmetry, and *antisymmetric*, where the values related by the centre have opposite signs. The s and d orbitals are centrosymmetric, the p and f orbitals antisymmetric.

More sophisticated pictures of orbitals than those in the figure show contours of the probability of finding an electron at a particular point. The sketches of orbitals elsewhere in this book are qualitatively of this type, representing sections through surfaces including most of the electron density. Strictly, these probability surfaces do not have signs, but the signs of the wave functions and the nodal planes can be added for clarity.

Further information: Most physical and inorganic chemistry textbooks discuss atomic orbitals. For a clear recent treatment, see Massey, A. G. (1990) *Main group chemistry*. Ellis Horwood, Chichester, pp. 37–49.

Box 6 Atomic orbitals (*contd.*)

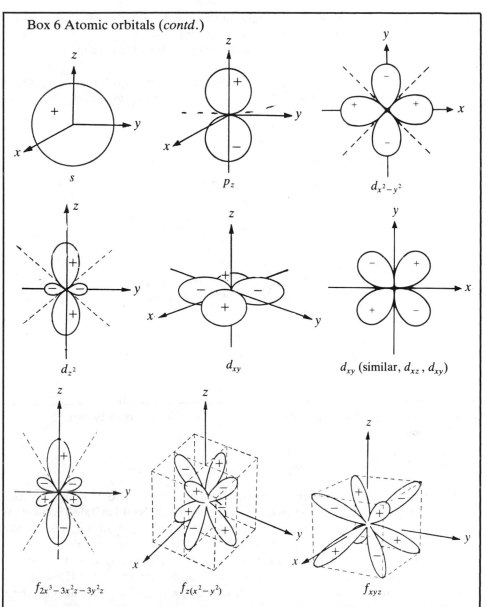

Fig. B6 — The angular wave functions for s, p, d, and f orbitals. (a) s (spherical); no nodes, (b) p_z; node in xy plane; p_x and p_y are similar, (c) d_{x2-y2}; nodes on diagonal planes, (d) d_{z2} (symmetrical by rotation about z axis; nodes are cones shown by dotted lines, (e) d_{xy} (two views); nodes on xy and xz planes; d_{yz} and d_{xz} are similar, (f) $f_{2z3-3x2z-3y2x}$ (symmetrical by rotation about z axis; nodes in xy plane and cones shown by dotted lines; two similar orbitals related to x and y axes, (g) $f_{z(x2-y2)}$; nodes in xy plane and 45° planes; two similar orbitals related to xz and yz planes, (h) f_{xyz}; nodes in xy, xz and yz planes.

(i) An *energy diagram* shows the energies of the atomic orbitals at each side and of
     the molecular orbitals in the centre, with linking tie-lines.
(ii) The orbitals are drawn to show the distribution of the electrons in space.

The two rules make it easy to construct diagram (i) for the interaction of two 1s
atomic orbitals on H-atoms (Fig. 5.2). The two atomic orbitals ψ_1 and ψ_2 combine to

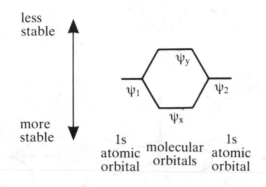

Fig. 5.2 — Energy diagram for the formation of molecular orbitals from the atomic orbitals of
two hydrogen atoms.

give two molecular orbitals. What linear combinations — produced by addition and
subtraction — are possible for two atomic orbitals? Obviously only two:

$$\psi_+ = \psi_1 + \psi_2 \quad \text{or} \quad \psi_- = \psi_1 - \psi_2$$
$$\text{addition} \qquad\qquad \text{subtraction}$$

Fig. 5.3 shows how these combinations change as two H-atoms approach each
other, and Fig. 5.4 gives precise pictures of the two orbitals. It is vital to remember
that each of the lower diagrams in Fig. 5.3 shows *one* molecular orbital, able to take
two electrons only (by the Pauli principle; Box 6). To keep the numerical values of
electron density correct, the wave functions (ψ_+, ψ_-, etc.) have to be multiplied by a
factor — the normalization factor — in this case approximately $1/\sqrt{2}$ (Box 7). For
simplicity, because they are unimportant except for detailed calculations, these
factors will normally be ignored.

5.3.1 Energies
The next step is to link the energy and the space picture of the molecular orbitals.
Which of the orbitals in Fig 5.3 will be of lower energy (more stable to removal of the
electron)?

Consider an electron at the point midway between the two nuclei (i.e. the species
H_2^+). It will attract both nuclei strongly, and will tend to hold the two nuclei together,
whose positive charges would otherwise repel each other. Thus, increasing the

---

Box 7 Normalization of orbitals

The probability of finding an electron at a given points is given by ψ^2. For any orbital, the summation of this over the whole of space must correspond to 100% probability of finding the electron, that is,

$$\int \psi^2 dv = 1.0$$

Molecular orbitals must be adjusted so that this is true, a process known as normalization. We can take the molecular orbital ψ_+ as an example, formed from the atomic orbitals ψ_1 and ψ_{12} with both coefficients 1.0.

$$\psi_+ = k(1.0\psi_1 + 1.0\psi_2)$$

Therefore $\int \psi_+^2 dv \ (=P) = 1.0$; k is the *normalization constant*.

$$P = \int k^2 (\psi_1^2 + \psi_2^2 + 2\psi_1\psi_2) dv$$

As the atomic orbitals ψ_1 and ψ_2 are already normalized, so

$$P = k^2(1.0 + 1.0) + k^2 \int 2\psi_1\psi_2) dv \qquad (B7.1)$$

The term, $\int \psi_1\psi_2) dv \ (=s)$, is known as the *overlap* of ψ_1 and ψ_2; it is built up from those portions of space where both orbitals have significant values. In the simplest calculations of molecular orbitals, s is assumed to be zero (e.g. in the Hückel treatment; section 9.8). For H$_2$, s is about 0.25.

From (B7.1), $P = k^2 (1.0 + 1.0 + 2s)$

i.e.

$$k = 1/\sqrt{2(1+s)} \text{ (or } 1/\sqrt{2} \text{ if overlap is neglected)}$$

For ψ_-, the anti-bonding orbital, $k = 1/\sqrt{2(1-s)}$.

The difference between these normalization constants affects the energies of ψ_+ and ψ_-, causing ψ_- to be raised above the mean energy of the original atomic orbitals.

Not all pairs of atomic orbitals can combine successfully. For example, in the interaction of a H(1s) orbital and a C(2p) orbital with its axis perpendicular to the C–H, line the overlap is precisely zero because of the symmetry of the orbitals (see Fig. 5.10). The same is true for any two orbitals of a single atom. Such pairs of orbitals are described as *orthogonal*.

---

electron density at this point will stabilize the molecule. In turn, the electron will be strongly held by the attraction of both nuclei. Of the two molecular orbitals formed from the H atomic orbitals, the *sum*, $\psi_1 + \psi_2$, has an increased density and the *difference*, $\psi_1 - \psi_2$, has a decreased density at the centre. In fact the second orbital has a nodal plane, and the phase of the wave function changes sign at the centre; the electron density is precisely zero. Thus an electron in ψ_+ is more stable than an electron in the atomic 1s orbital, and one in ψ_- less stable.

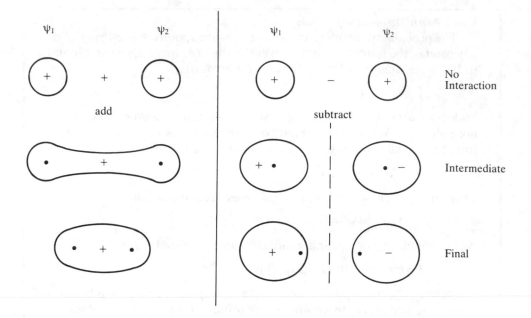

Fig. 5.3 — Electron distribution during the formation of bonding and anti-bonding orbitals from the atomic orbitals of two hydrogen atoms.

5.3.2   Bonding and antibonding orbitals

A very useful property of molecular orbitals is that the general arrangement of their energy levels does not depend on which individual orbitals are filled. The LCAO approach to bonding is therefore particularly helpful if the number of electrons changes. We can use Fig. 5.2 to consider the bonding between two hydrogen nuclei combined with one, two, three, or four electrons.

Consider the simplest position, two hydrogen nuclei with just one electron. This electron will occupy the lowest orbital (Fig. 5.5) and will be more strongly held than in a free hydrogen atom, actually by $256 \, \text{kJ mol}^{-1}$. This gain in stability is the *bonding* energy of the H_2^+ ion; to break the ion up into $H \cdot + H^+$, this energy must be provided.

When this one electron is put into ψ_+ $(=\psi_1+\psi_2)$, the two hydrogen nuclei are bonded together. This orbital is therefore called a *bonding orbital*. In contrast, the combination of two nuclei and one electron in ψ_- $(=\psi_1-\psi_2)$ is less stable than separate $H \cdot$ and H^+. This orbital is therefore called an *antibonding orbital*. Molecules also often have *non-bonding molecular orbitals* containing electrons with about the same energy as in the free atoms (e.g. section 5.5.4). In this book, the orbitals are labelled ψ_b, ψ_{ab}, and ψ_{nb} respectively.†

A second electron added to H_2^+ will go into the most stable available orbital. This is the bonding orbital ψ_b (i.e. ψ_+), though to enter it, the electron must have the

† This identifies the orbitals more clearly than the simpler notation ψ and ψ^* which is often used for bonding and antibonding orbitals. In complicated molecular orbital diagrams, the orbitals are numbered in order of their energy.

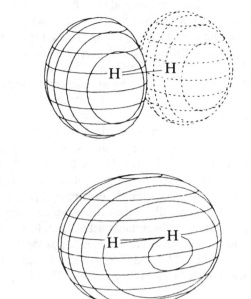

(a)

(b)

Fig. 5.4 — Three-dimensional probability contours for the formation of bonding (a) and anti-bonding (b) orbitals by two hydrogen atoms. Reproduced with permission from Jorgensen, W. L. & Salem, L. (1973) *The organic chemist's book of orbitals*. Academic Press, New York, Fig.1.

Filling of H_2 molecular orbitals

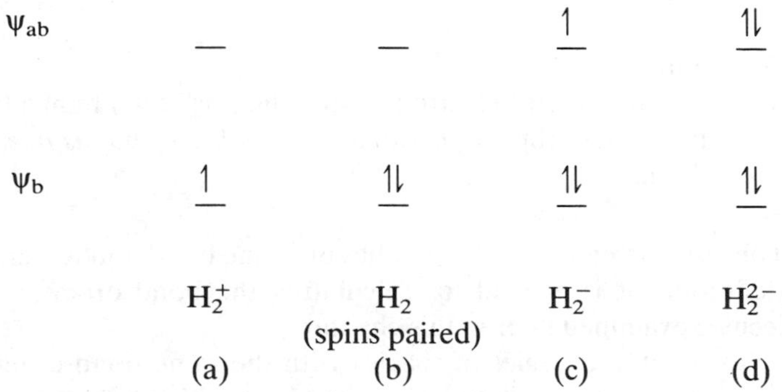

Ψ_{ab}	—	—	↑	↑↓
Ψ_b	↑	↑↓	↑↓	↑↓
	H_2^+	H_2 (spins paired)	H_2^-	H_2^{2-}
	(a)	(b)	(c)	(d)

Fig. 5.5 — Filling of the molecular orbitals in H_2 by electrons.

opposite spin to the first electron (obeying the Pauli principle). The result is the H_2 molecule, and the total stabilization compared to two isolated H-atoms is now -432 kJ mol^{-1}. (Note the minus sign, indicating an increase in stability.) This value is not twice that for H_+^2 because the two electrons in the orbital repel each other, reducing the stability of the system. The difference between 2×256 and 432 (80 kJ mol^{-1}) gives a minimum value for the repulsive energy. This calculation assumes that the orbital energies are unchanged; in reality, the greater attraction of the two bonding electrons for the nuclei brings the nuclei closer together in H_2 (74.1 pm) than in H_2^+ (105.2 pm), increasing the splitting between the orbitals.

   If the spin of the second electron is parallel to the first, it has to enter the antibonding orbital and its loss in stability cancels out that gained from the first electron. Adding a third electron to H_2 gives H_-^2. Obviously, this electron goes into the antibonding orbital, decreasing the stability of the molecule, but H_-^2 is still stable; though the energy has not been well-determined, it is believed to be between -100 and -200 kJ mol^{-1}. The smaller value seems more likely, as He_+^2, with the same number of electrons but twice the nuclear attraction, has an energy of -228 kJ mol^{-1}. Finally, with a fourth electron all the bonding is destroyed. H_2^{2-} and He_2 have no advantage compared to $2H^-$ and $2He$ — in fact a disadvantage because of the relative destabilization of the antibonding orbital (Box 7).

5.3.3   Bond order and bond energy

On a very simple bonding model, a single bond contains a pair of electrons. We therefore calculate the *bond order* as the net number of *bonding* pairs of electrons.

   For the molecules and ions containing two H-atoms:

   H_2^+ has 1 bonding electron, i.e. 1/2 a pair
        Bond order $= 1/2$
   H^2 has two bonding electrons (one pair),
        Bond order $= 1$
   H_2^- has two bonding electrons, but the one antibonding electron cancels one of
        these; net 1 bonding electron,
           Bond order $= 1/2$

As a formal definition:
   Bond order $= 1/2$ [(number of electrons in bonding orbitals) minus (number of electrons in antibonding orbitals)] divided by (*number of bonds over which the electrons are delocalized*).

The italicized phrase is irrelevant to H_2 as it has only one bond, but it is an important part of the definition; it is needed for calculating the bond orders of the more complex molecules examined later in this chapter.

   We cannot expect the energies of species with the same bond order to be the same. This appears even in comparing H_2^- and H_2^+, and is much more marked as soon as different sorts of atoms are involved. As an example, in O_2^{2-}, the bond order is 1 (section 5.9.1), and the net energy relative to two O^- is -204 kJ mol^{-1} (compared to -432 kJ mol^{-1} in H_2). This difference in the numerical values of the energy is not a weakness of the classification by bond order, but a strength. For consideration of their bonding, it is more useful to correlate O_2^{2-} and H_2, each with bond orders of 1, than to compare their bond energies.

5.4 MULTI-ATOM MOLECULES

We have looked at H_2 in considerable detail, despite its being the simplest possible molecule. Fortunately, we can apply the same principles straightforwardly to more complicated molecules. We need to extend the molecular orbital bonding model in two ways:

(a) to molecules with more than two atoms,
(b) to molecules with multiple bonds.

5.4.1 Hybridization — CH$_4$

The methane molecule has a regular tetrahedral shape, with all C−H distances equal (109.4 pm). Looking at the ground electronic state of the carbon atom, this regularity is surprising (Fig. 5.6a). Only the two p-electrons seem able to form bonds. However, the energy difference between the 2s and 2p levels is not great, and one electron can easily be promoted from the 2s to the 2p level, to give four unpaired electrons (Fig. 5.6b). These electrons can now form four bonds, but these bonds

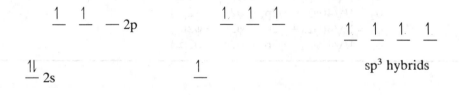

(a) Carbon atom: ground state   (b) One electron promoted   (c) Hybridised

Fig. 5.6 — Carbon atom: (a) ground state (omitting 1s orbital) (b) promotion of a 2s electron (c) formation of sp^3 hybrid orbitals.

would still not be equivalent. In a further stage, the four different orbitals (2s + three 2p) can be mixed together to give four identical orbitals (Fig. 5.6c). This is known as *hybridization* (Box 8), and the resulting orbitals are *hybrid atomic orbitals*, in this case sp^3 orbitals.

Fig. 5.7 shows the shape and arrangement of these orbitals. Apart from their

Box 8   Hybrid orbitals
    Hybrid orbitals are formed by the mixing together of two or more atomic
orbitals of one atom. They are named according to the atomic orbitals from
which they are formed. Their shapes are shown in the Figure for the
examples most often encountered.

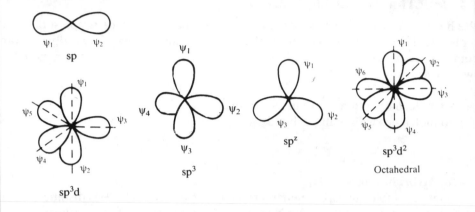

Fig. B8 — Shapes of hybrid orbitals (a) sp, (b) sp^2, (c) sp^3, (d) sp^3d, (e) sp^3d^2.

    The mathematical forms for hybrid orbitals involving only s and p
orbitals are rather simple and symmetrical. As an example, the four sp^3
orbitals are described by the following equations (including the normalization constants $(1/\sqrt{4})$:

$$\psi_1 = 1/\sqrt{4}(\psi 2s + \psi 2p_x + \psi 2p_y + \psi 2p_z)$$
$$\psi_2 = 1/\sqrt{4}(\psi 2s - \psi 2p_x - \psi 2p_y + \psi 2p_z)$$
$$\psi_3 = 1/\sqrt{4}(\psi 2s + \psi 2p_x - \psi 2p_y - \psi 2p_z)$$
$$\psi_4 = 1/\sqrt{4}(\psi 2s - \psi 2p_x + \psi 2p_y - \psi 2p_z)$$

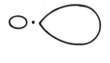

(a)

(b)

Fig. 5.7 — (a) One sp_3 hybrid orbital, (b) the set of four such orbitals (minor lobes omitted for
clarity).

direction, pointing from the centre to the corners of a tetrahedron, their most important feature is their shape. Atomic orbitals are all *centrosymmetric* or *antisymmetric* (Box 6). In contrast, hybrid orbitals are formed by combining centro-symmetric and antisymmetric orbitals. They therefore have a main lobe pointing in *one direction* away from the atomic nucleus, very suitable for overlapping with an orbital on another atom.

Note the difference between hybrid atomic orbitals and molecular orbitals. Both are formed by combining n atomic orbitals to give n product orbitals, but the hybrid orbitals are equal in energy; the molecular orbitals are unequal and involve more than one nucleus.

5.4.2 C−H bonds in CH$_4$

Formation of the molecular orbitals for a C−H bond follows exactly the same procedure as for H$_2$, starting from an sp^3 orbital on C and a 1s orbital on H. Fig. 5.8

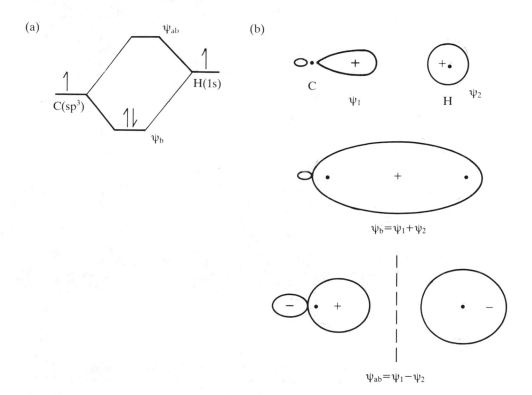

Fig. 5.8 — The formation of a C−H bond (a) energy diagram, (b) contours of electron probability.

represents the process in energy and space diagrams. A difference between this and the formation of an H−H orbital is that the C(sp^3) and H(1s) orbital do not have the same energy. However, as long as the two atomic orbitals are not very different in energy, they will combine satisfactorily.

Each of the four sp^3 orbitals and each of the 1s orbitals starts with one electron. Thus, each of the bonding molecular orbitals in CH_4 contains a pair of electrons, as shown in the molecular orbital energy diagram for the complete molecule (Fig. 5.9a).

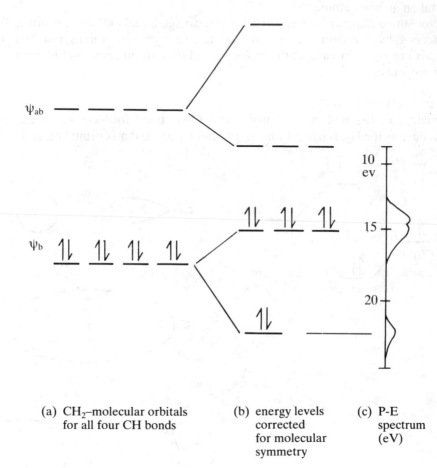

(a) CH_2–molecular orbitals   (b) energy levels   (c) P-E
    for all four CH bonds   corrected   spectrum
                                         for molecular   (eV)
                                         symmetry

Fig. 5.9 — CH_4 (a) Molecular orbital energy levels for all C–H bonds, (b) Energy levels corrected for molecular symmetry, (c) the corresponding photoelectron spectrum.

Each bond therefore has a bond order 1.0.

Inevitably, such diagrams are oversimplifications. A more precise treatment shows that the four identical ψ_b orbitals will be mixed together, producing one orbital of lower energy and three of higher energy (Fig. 5.9b); the ψ_{ab} orbitals interact similarly. The photoelectron spectrum of methane (Fig. 5.9c) confirms the pattern. In most molecules the interaction between the molecular orbitals does not affect the overall bonding, and can be ignored. It is important in metal complexes and they therefore need a more precise treatment.†

5.5 MULTIPLE BONDING

Simple pictures of ethene (ethylene) (I) and ethyne (acetylene) (II) show them with double and triple bonds. It is obvious that they need more complicated molecular orbital descriptions than does H_2. The key feature of these molecules is their geometry: C_2H_4 is planar and C_2H_2 linear.

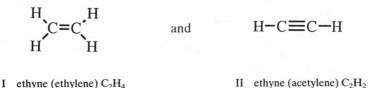

	and	

I   ethyne (ethylene) C_2H_4 II   ethyne (acetylene) C_2H_2

5.5.1 C_2H_4
We need first to consider the effect of this molecular geometry on the interaction of the atomic orbitals. What happens if we attempt to form molecular orbitals from a 1s orbital on H and the p-orbital on C which has its lobes *perpendicular to the molecular plane* (say the p_z orbital) (Fig. 5.10)?

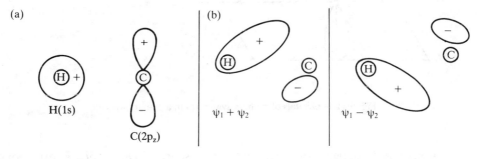

(a)

H(1s)   C(2p_z)

(b) $\psi_1 + \psi_2$ $\psi_1 - \psi_2$

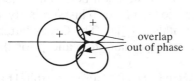

(c)

overlap
out of phase

Fig. 5.10 — Overlap between H_{1s} and C_{pz} orbitals in C_2H_4 (a) the atomic orbitals, (b) possible molecular orbitals, (c) areas of overlap between atomic orbitals.

† A more formal approach involving the construction of what are known as *terminal atom symmetry orbitals* (TASO) can be found in Purcell & Kotz (1977), Chapter 4. See also Kettle, (1986).

Two molecular orbitals, $\psi_+ = \psi_1 + \psi_2$ and $\psi_- = \psi_1 - \psi_2$, can be constructed, as before. What about their energies? Because of the geometry of the orbitals, whatever overlap occurs in phase (increasing stability) on one side is exactly cancelled by the out-of-phase overlap on the other side! These two orbitals cannot combine to form molecular orbitals; they are orthogonal (Box 7).

Because the C_2H_4 molecule is planar, we can divide the atomic orbitals into two sets which interact only with other orbitals within the same set. Thus, the $C(2p_z)$ and $H(1s)$ orbitals must be in different sets (Fig. 5.11).†

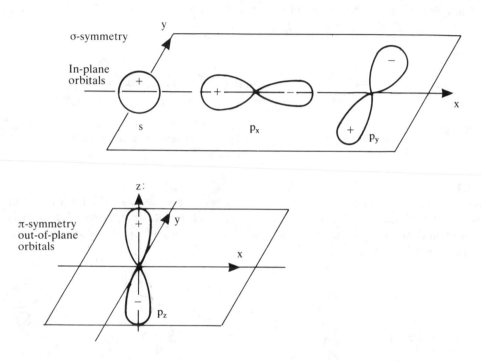

Fig. 5.11 — Orbitals of σ and π symmetry in a planar molecule.

The separation depends on the absence or presence of a *nodal plane* corresponding to the plane of the molecule. This is similar to the distinction between s and p atomic orbitals, respectively without and with a nodal plane. This similarity is reflected in the names given to the two types of orbitals in C_2H_4. The Greek letters σ(sigma) and π(pi) label the molecular orbitals, corresponding to s and p for the atomic orbitals.‡

Fig. 5.11 shows the orbitals belonging to each set in C_2H_4. Because a π-orbital has

† The difference between this lack of interaction and the formation of hybrid orbitals from a centro-symmetric and an antisymmetric orbital is that the latter are formed from s and p orbitals on *the same* atom.

‡ δ(delta) molecular orbitals, with two nodal planes, also exist, corresponding to d-orbitals. These δ orbitals are important in some classes of transition metal complexes, such as $[Re_2Cl_8]^{2-}$.

a nodal plane, only p (or d or f) orbitals (with nodal planes themselves) can be of π-symmetry. s orbitals must be of σ-symmetry, while p orbitals may be, if their nodal plane does not coincide with the molecular plane. *Note that a p orbital is NOT automatically of π-symmetry.*

5.5.1.1   σ-molecular orbitals of C_2H_4

We can now easily construct the bonding scheme for C_2H_4, considering first the σ- and then the π-orbitals (Figs 5.12–13). As with CH_4, the first stage is hybridization.

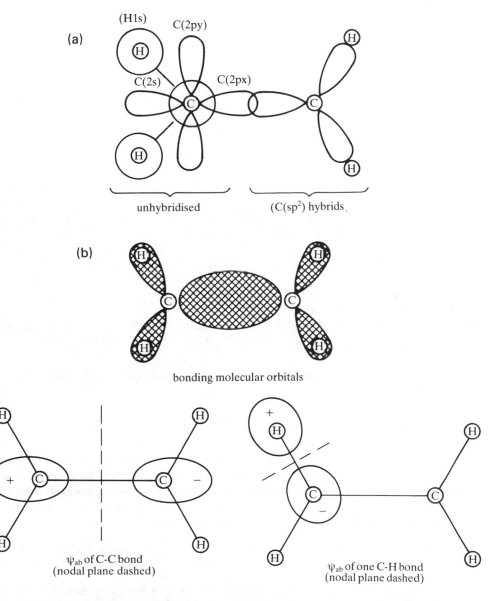

Fig. 5.12 — σ orbitals in C_2H_4 (a) atomic orbitals unhybridized and hybridized (b) bonding and antibonding molecular orbitals.

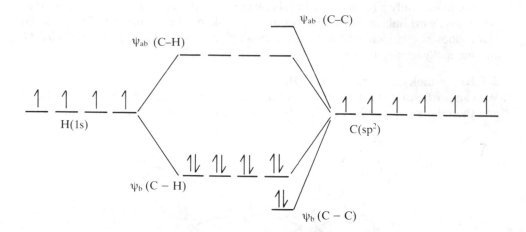

Fig. 5.12 — σ orbitals in C_2H_4 (c) energy diagram.

Each C-atom has an s and two p orbitals of σ-symmetry. These form sp^2 hybrids, pointing to the corners of a planar triangle (Box 8; section 5.6.1). These overlap either with another sp^2 orbital on the other carbon, or a 1s orbital on H. Each pair produces a bonding and an antibonding molecular orbital, exactly as with H_2. Finally, each of the five bonding σ orbitals is occupied by a pair of electrons, producing a bond order of 1.0 between each adjoining pair of atoms.

5.5.1.2   π-molecular orbitals
Formation of the π-molecular orbitals is also straightforward (Fig. 5.13). With just two π-symmetry atomic orbitals (ψ_1 and ψ_2), the two molecular orbitals are:

$$\psi_b = \psi_1 + \psi_2$$
$$\text{and} \quad \psi_{ab} = \psi_1 - \psi_2$$

C_2H_4 contains 12 valence electrons (C: 2×4; H: 4×1), of which 10 are in σ-bonding molecular orbitals. Thus, the last two electrons occupy the π_b orbital. As this is located between the two central atoms, the electron pair contributes one additional bonding pair to this bond, giving it a total bond order of 2.0, as we expect from the simple bond drawing (I).

5.5.2   Orbitals of C_2H_2
C_2H_2 (II) is a linear molecule. The same reasoning that led us to consider the p_z orbitals separately in C_2H_4 therefore applies to both p_y and p_z orbitals in C_2H_2. These

(a) $C(p_z)$

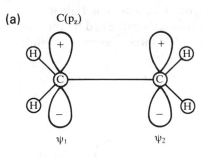

(b)

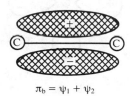

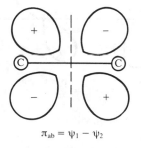

(c)

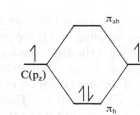

(d)

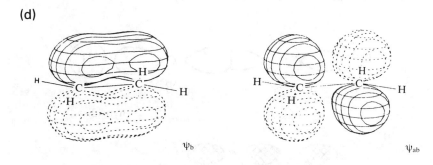

Fig. 5.13 — π orbitals in C_2H_4 (a) atomic orbitals, (b) molecular orbitals, (c) energy diagram, (d) calculated probability contours. Source for (d): as Fig. 5.4. (Fig. 18).

both have nodal planes perpendicular to the *molecular axis* (rather than lying in the molecular plane). Furthermore, a p_y and a p_z orbital cannot combine together to give molecular orbitals, as any in-phase overlap is exactly cancelled by the out-of-phase overlap (Fig. 5.14). Thus, this linear molecule has three classes of orbital:

σ: H, 1s; C, 2s, 2p$_x$
π_y: C, 2p$_y$
π_z: C. 2p$_z$

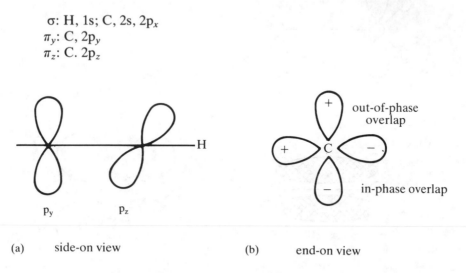

(a)   side-on view   (b)   end-on view

Fig. 5.14 — Overlap of p_y and p_z orbitals in C_2H_2. (a)side-on view, (b) end-on view.

The two σ orbitals on carbon are hybridized sp, producing a linear arrangement. These overlap with 1s on H, giving a total of three bonding and three antibonding molecular orbitals (Fig. 5.15); each of the bonding orbitals contains a pair of electrons. The π_y orbitals produce one bonding and one antibonding orbital (identical to those in Fig. 5.13f), as do the π_z orbitals (energy diagram in Fig. 5.16). Each of these π-bonding orbitals also contains one pair of electrons. Thus in all, the C–C bond has three bonding pairs (σ+two π), giving a bond order of 3.0, while each C–H bond, with one σ-bonding pair has a bond order of 1.0. The C–C bond can be compared to the triple bond of N_2 (section 5.9.1).

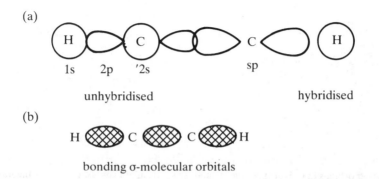

Fig. 5.15 — σ orbitals in C_2H_2 (a) atomic orbitals unhybridized and hybridized, (b) bonding molecular orbitals.

(c)

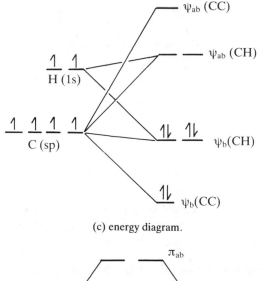

(c) energy diagram.

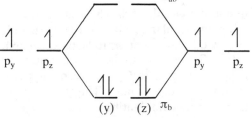

Fig. 5.16 — Energy diagram for π orbitals in C_2H_2.

5.5.3 Two extended molecules

The molecules (III) and (IV–V) each have two double bonds. In C_5H_8 the terminal atoms can rotate so that the carbon skeleton is planar, but the presence of the two central H-atoms (marked) means that strictly it is non-planar. Therefore, all atomic orbitals are of σ-symmetry. In reality, the interaction between the out-of-plane 1s orbitals of the central H-atoms and the p_z orbitals of the C-atoms will be small, and the molecule can be treated as having two separate π-bonds.

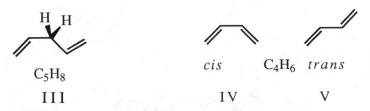

C₄H₆ (IV–V) is more interesting. It is normally planar, either *cis* (IV) or *trans* (V), as shown. Unlike (III), the two p_z orbitals in the centre (perpendicular to the molecular plane) (Fig. 5.17) do overlap — as strongly as the pairs at each end! Therefore the π-molecular orbitals have to be formed from all four p_z orbitals, not from two separate pairs. Such adjacent double bonds are called *conjugated*. Their

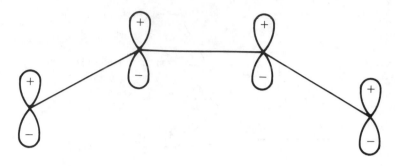

Fig. 5.17 — π atomic orbitals (p_z) in C_4H_6.

properties are very different from the simple combination of two double bonds that (IV–V) suggests. These and other conjugated organic molecules are looked at in more detail in Chapter 10.

5.5.4   Formation of σ- and π-molecular orbitals: a summary

Fig. 5.18 summarizes how molecular orbitals are constructed, using the method described here, with localized σ-molecular orbitals, based on hybrid atomic orbitals, and delocalised π orbitals. The figure applies the method to methanal (formaldehyde) $H_2C{=}O$. This is very similar to C_2H_4, except that two of the oxygen sp^2 hybrid orbitals overlap with no other σ orbitals. They therefore remain unchanged in energy, and are neither bonding nor antibonding. They are described as *nonbonding orbitals* (ψ_{nb}). In considering the addition of electrons to the molecular orbitals, they should therefore be filled after the π_b orbitals. In reality, in all the molecules likely to be considered, they are occupied, and it is therefore convenient to include their electrons at step 5.

Stage	Action	Illustration
1	Assume (or choose for study) the molecular geometry	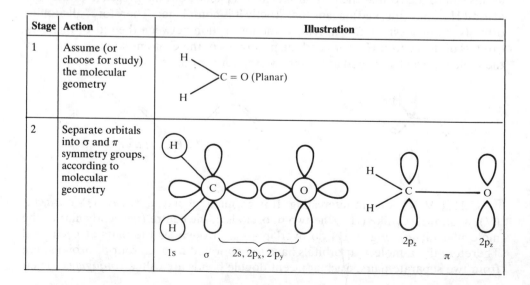
2	Separate orbitals into σ and π symmetry groups, according to molecular geometry	

Stage	Action	Illustration
3	Convert σ-symmetry orbitals to hybrid orbitals	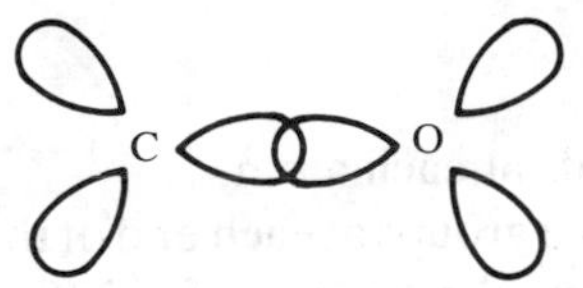
4	Form ψ_b and ψ_{ab} molecular orbitals for localized σ-bonds. Identify any lone pair (ψ_{ab}) orbitals	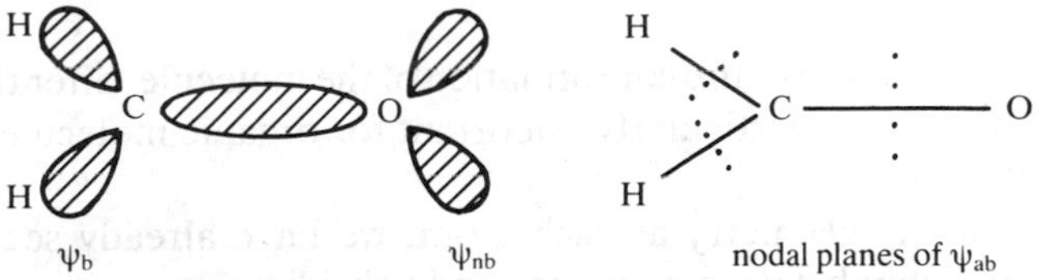 ψ_b ψ_{nb} nodal planes of ψ_{ab}
5	Fill ψ_b, ψ_{nb} orbitals with pairs of electrons	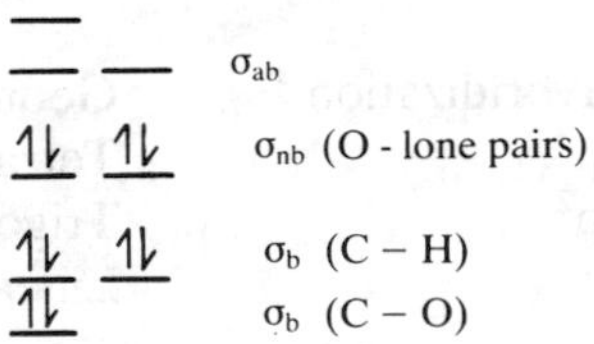 σ_{ab} σ_{nb} (O - lone pairs) σ_b (C – H) σ_b (C – O)
6	Form π-molecular orbitals	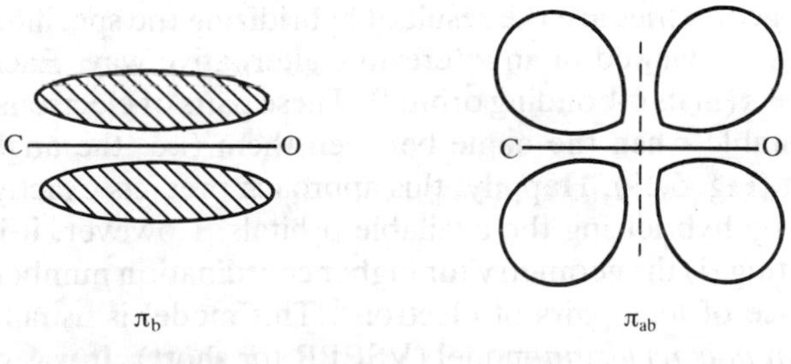 π_b π_{ab}
7	Fill π-molecular orbitals with available electrons	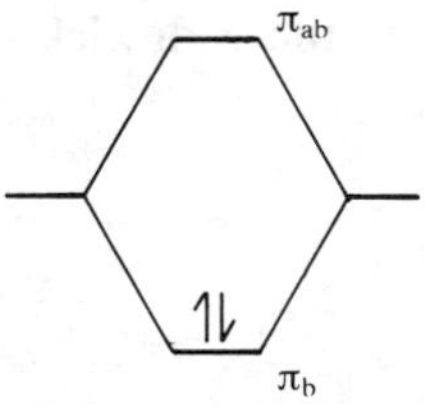 π_{ab} π_b
8	Calculate Bond Order:	C – H bonds: 1.0 C = O bond: 2.0

Fig. 5.18 — Construction of molecular orbitals molecules with π-bonds: formaldehyde (methanal).

5.6   SHAPES OF COVALENT MOLECULES

The shape of a molecule is determined by the most favourable values of

(1) the angles between the bonds at each atom,
(2) the relative orientations of the groups at each end of the bonds (usually described by the torsion angles, showing the amount of twist in each bond).

Point (2) concerns the conformation of the molecule, after the bond angles have been specified. It is particularly important for organic molecules (e.g. Bassindale, 1984, Testa, 1979).

For the geometry at each atom, we have already seen three examples of the relationship between geometry and hybridization:

Molecule	Hybridization	Geometry
CH_4	sp^3	Tetrahedral
C_2H_4	sp^2	Trigonal planar†
C_2H_2	sp	Linear

These geometries are the result of hybridizing the specified sets of orbitals, but they can also be viewed in an interesting alternative way. Each bond contains a pair of electrons (in its σ-bonding orbital). These pairs of electrons repel each other, and are most stable when the angle between them (i.e. the angle between the bonds) is largest (Fig. 5.19). Happily, this approach predicts exactly the same geometry as is found by hybridizing the available orbitals. However, it is particularly valuable in predicting (i) the geometry for higher coordination numbers [see Box 4], and (ii) the influence of lone pairs of electrons. This model is usually called the *valence shell electron pair repulsion* model (VSEPR for short). It was originally proposed by N. V. Sidgwick and H. M. Powell, and later developed by R. J. Gillespie and R. S. Nyholm. It is therefore also known as the Sidgwick–Powell and the Gillespie–Nyholm model.

5.6.1   Geometry predictions

The geometries predicted by VSEPR for any number of electron pairs from 2 to 7 are shown in Fig. 5.20, with the corresponding orbitals used in hybridization (cf. Box 8).

---

† 'Trigonal'=*three-sided* i.e. triangular. Trigonal is used in parallel to pentagonal, etc., though *square* is usually preferred to tetragonal, because the latter has a special significance as the name for a crystal system (section 2.1.1).

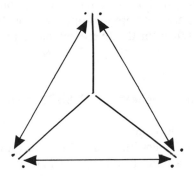

Fig. 5.19 — Repulsion between three pairs of electrons. A trigonal planar arrangement minimises these repulsions.

No. of electron pairs	Geometry	Drawing	Hybridisation
2	linear		sp
3	trigonal		sp^2
3	planar		
4	tetrahedral		sp^3
5	trigonal		sp^3d
5	bipyramid		
6	octahedral		sp^3d^2
7	pentagonal byramid		sp^3d^3

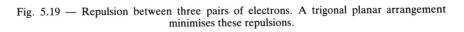

Fig. 5.20 — Preferred geometries for different numbers of electron pairs.

(No covalent compounds with 8 or more pairs of electrons are known.) With four pairs, the s and p orbitals are used up, and one, two, or three d orbitals have to be included. These are chosen from the five available orbitals, to produce the right geometries.

5.6.1.1   Geometry of PCl₅

We can now look at a simple application of VSEPR, to the molecule PCl_5:

(a)   P has 5 valence-shell electrons   5
(b)   Each Cl contributes 1 electron to a P-Cl bond (adding to the total
      number of electrons around P; the other Cl electrons are not shared
      with P)   5x1
(c)   Total number of electrons: 10=5 pairs   10
Deduce
(d)   Geometry: trigonal bipyramid

This fits exactly with the structure of the molecule in the gas phase (Fig. 5.21).

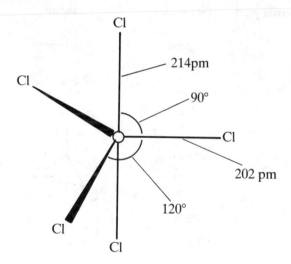

Fig. 5.21 — Experimental geometry for PCl_5.

The slight difference between axial and equatorial bond lengths is interesting. It shows firstly that the trigonal bipyramid (as also the pentagonal bipyramid) is of lower symmetry than the tetrahedron and octahedron, which have all their bonds equal. The slightly longer axial bonds may be a result of the electron repulsion. These electron pairs experience greater repulsion than the equatorial pairs (as explained below).

5.6.2   Lone pairs — NH₃

The great strength of the VSEPR approach lies in its ability to deal with lone pairs. Consider a simple example, NH_3 in the same way as PCl_5:

(a) N has 5 valence electrons   5
(b) Each H contributes 1 electron to a N–H bond   3x1
(c) Total electrons: 8=4 pairs   8
Deduce
(d) Predicted geometry: tetrahedral but only 3 electron pairs are in N–H
     bonds; the fourth is a lone pair.

   Thus, the observable geometry is pyramidal (Fig. 5.22) because the experimental

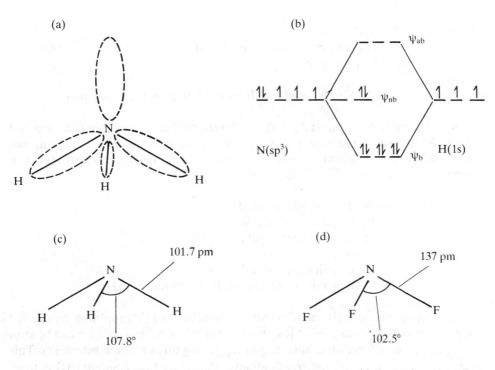

Fig. 5.22 — NH_3 (a) ideal structure, (b) molecular orbital energy levels, (c) observed geometry, (d) NF_3 observed geometry.

techniques show only the atomic positions. We deduce that the lone pair occupies the
fourth position of the tetrahedron. The simple energy level diagram (Fig. 5.22b)
shows three low-energy molecular orbitals and one higher, corresponding to the
N–H bonds and the lone pair. If the effects of symmetry are included, the three
equivalent bonding orbitals split to give one of somewhat lower and two of somewhat
higher energy.
   The VSEPR model also predicts the deviations from regularity of the observed
structure of NH_3. The electron pairs in the N–H bonds are attracted by both N and H
atoms, but the lone pair is attracted by only the N-atom. It is therefore closer to N
and takes up more space around N. The result is that the angle between two H–N

bonds is less than the value for an ideal tetrahedron: 107.8° instead of 109.4°. The structure of NF_3 shows this effect on a larger scale. F-atoms attract electrons more strongly than H, because of their greater electronegativity. The more tightly bound electrons in the N–F bonds take up even less space than those in the N–H bonds. Thus the F–N–F angle (102.5°) is smaller than the H–N–H angle.

5.6.3   Lone pairs in trigonal bipyramids — SF_4

Predicting the structure of SF_4 is slightly more tricky. The calculation of electron pairs is as before:

(a)	S has 6 valence shell electrons	6
(b)	Each F contributes 1 electron to an S–F bond	$\dfrac{4 \times 1}{10}$
(c)	Total electrons: 10=5 pairs	

Deduce

(d)   Geometry: trigonal bipyramid with four S–F bonds and one lone pair.

Because the axial and equatorial positions of a trigonal bipyramid are different, we have to decide where the lone pair is located. We can do this by considering the repulsions between the electron pairs, as measured by the angles between them. The alternatives are shown in Fig. 5.23.

    (i) Equatorial: the lone pair is repelled by
                2 bond pairs (axial) at 90°
            +2 bond pairs (equatorial) at 120°
   (ii) Axial: the lone pair is repelled by
                3 bond pairs (equatorial) at 90°
        (+1 pair at 180° which can probably be neglected)

Thus the change between (i) and (ii) involves replacing $2 \times 120°$ repulsions by one 90° repulsion. It is difficult to predict just what effect this will have, but we can bear in mind that repulsive forces are short-range (e.g. falling off as r^6 for a distance r). This suggests that the 90° repulsion will dominate, indicating that geometry (i) is most likely; experiment (Fig. 5.23b) confirms this.† The bond angles also show the distortions to the ideal structure caused by the lone-pair/bonded-pair repulsions.

5.6.4   The effect of π-bonds: $POCl_3$

The presence of a π-bond is an extra factor in the VSEPR approach. The double bond involves two pairs of electrons — one in the σ- and one in the π-bonding molecular orbital. However, both pairs point in the same direction, and so for VSEPR predictions count only as one. Taking $POCl_3$ (VI) as an example, a simple bond diagram shows a double P=O bond and three single P–Cl bonds. The electron-counting procedure is the same as before, apart from one extra step:

---

† Correctly, this geometry should be described in terms of the atomic positions only, as a distorted tetrahedron. For SF_4, this is very uninformative, so that it is generally described as having trigonal bipyramidal geometry with one vacant position (or as ψ-trigonal bipyramidal, meaning pseudo-trigonal bipyramidal).

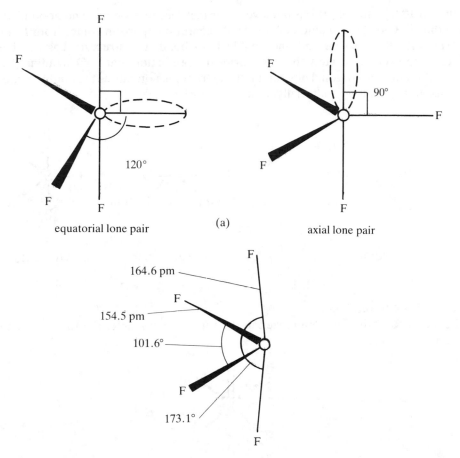

equatorial lone pair   (a)   axial lone pair

(b) observed

Fig. 5.23 — (a) Alternative geometries for four bond and one lone pair in SF$_4$, (b) observed geometry.

(a)   P has 5 valence electrons   5
(b)   O contributes 2 electrons and each Cl   1×2
      1 electron in P–O and P–Cl bonds   3x1
(c)   Total electrons; 10=5 pairs   ――
                                                                   10
(d)   Deduct pairs of electrons in π-orbitals, i.e. one pair in the P=O bond
      Net electrons: 4 pairs
Deduce
(e)   Geometry: tetrahedral.

Because they contain two pairs of electrons, double bonds exert rather stronger repulsions than do single bonds. Thus the geometry of POCl$_3$ shows a slight increase in the O–P–Cl angle and a decrease in the Cl–P–Cl angle from the ideal tetrahedral value to 103.3°. In fact, this is not a very fair test, because we have no good idea of whether P–O or P–Cl would be bulkier. A clearer comparison comes from H_3PO_4 (VII), with three single and one double P–O bond; the structure has not been determined very accurately, but it confirms the prediction that P=O is bulkier than P–O. Similarly in a trigonal bipyramid, the extra repulsion causes the double bonded groups to be located equatorially, as in SOF_4 (VIII).

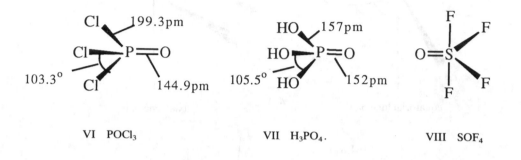

| VI   POCl₃ | VII   H₃PO₄. | VIII   SOF₄ |

5.6.5   VSEPR summary
The VSEPR method is summarized in the following example, the rather complex case of $XeOF_4$ (IX):

IX   XeOF₄.

I — Calculate number of valence shell electron pairs

(a)   Valence electrons of central atom   Xe   8
(b)   Contribution from electrons in bonds from   F   4×1
      outer atoms   O   1×2
                                                                                   Total 14
(c)   Deduct any electrons in π-bonds   Less 2 els   =12 els

II—   Predict geometry from net number of
      electron pairs   6 pairs=octahedral
III — If alternatives possible, place lone pairs   (a)   lone pair, any location
      and/or π-bonded groups by considering
      repulsions

                                                         (b)   Xe=O as far away as
                                                               possible, i.e. trans to
                                                               lone pair

IV —   Consider distortions caused by lone pair/   Close balance between
       π-bond repulsions   lone pair and π-bond re-
                                                         pulsions, so little distor-
                                                         tion

In fact, the determination of the structure of $XeOF_4$ by microwave spectroscopy is of
rather low precision. The overall structure is as predicted, and the determined value
for the O–Xe–F angle is 91.8°, suggesting that the π-bond repulsions are slightly the
greater. However, the error in this value is 0.5° (one standard deviation), so the
conclusion is not very firm.

5.7 COVALENT RADII

To establish standard radii for covalent bonds is even more important than for the
other bond types, because covalent bonds vary far more in length and strength.
These variations can best be studied by comparing individual bond lengths with well-
established standard single bond distances. Furthermore, some generally quoted
radii are not very satisfactory, leading to wild proposals for their correction.[†]
    A test for self-consistent covalent radii can be carried out, much as for ionic radii
(Box 9). The result shows that covalent radii should reproduce observed interatomic

---

Box 9   Consistency of covalent radii
    Subtraction of pairs of bond lengths can be used to examine the
consistency of covalent radii. A simple example shows that the difference
between the covalent radii of N and C is constant to about 1% of the N–C
bond distance

NH_2–CH_3	147.4		CH_3–CH_3	153.4
NH_3	101.5		CH_4	109.3
	———			———
difference	45.9		difference	43.8

---

distances to a precision of about 2 pm. Determining radii is much easier than for ionic
compounds, because identical covalent atoms can form bonds to each other. The
simplest values for covalent radii are therefore *half the distances between identical
singly bonded atoms*. Unfortunately, these radii do not give very consistent results,

---

[†] Notably the Schomaker–Stevenson correction. This was intended to improve agreement between
observed and calculated distances, but it has been shown to have the opposite effect (Wells 1984, p. 237).
Any predictions of distances using it should be examined very critically.

particularly for the most electronegative atoms (sections 5.8.2 and 9.5). Instead, the most useful radii are those derived by *subtracting the radius of carbon from C–X distances*. The usefulness of these carbon-based radii seems to relate to three negative properties: carbon has no lone pairs, no empty p- or d-orbitals, and is not of high electronegativity. Each of these can perturb observed bond distances (sections 7.3, 9.4–5).

The radius of carbon itself can be calculated from the bond distance in saturated hydrocarbon chains as 76.7 pm. Ideally, all single C–C bonds between sp^3 hybridized atoms should be identical, but a detailed comparison (Allen *et al.* 1987) shows small systematic variations. These are probably associated with changes in the effective electron-attracting power of the various groups attached to the bonded carbon atoms.

The calculated carbon-based radii are listed in Appendix A.1 and shown graphically in Fig. 5.24. The trends are characteristic of all scales of atomic size: (i) a

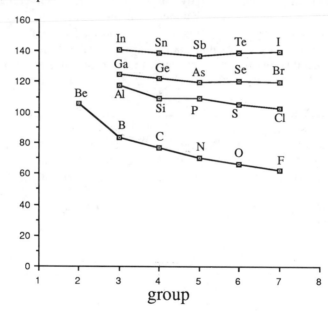

Fig. 5.24 — Trends of covalent radius by group.

decrease along a period as the addition of extra electrons does not produce much extra screening, (ii) an increase down a group following the addition of electrons to a new valence shell.

5.7.1 Multiple bonds
Knowing the bond order in compounds like C_2H_2 and C_2H_4 (section 5.5), we can use their C–C distances to calculate double and triple bond radii for carbon (66.1 and

59.1 pm). These in turn provide O=, N= and N≡ radii (Appendix A.1); from these, we can plot curves relating bond order and radius (Fig. 5.25). For short bonds, these

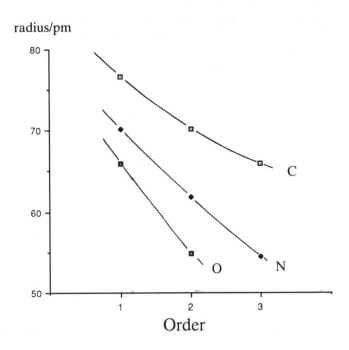

Fig. 5.25 — Multiple bond radii for C, N, O.

curves are well-defined, because the bond orders are easily calculated. Bonds weaker than 1.0 do occur, though they are very unusual for carbon, and the calculation of their bond order requires assumptions about their bonding (section 7.3.3). Additionally, we have tentative evidence that the bond order is very small (close to zero) at the normal distance for intermolecular contacts, about 150 pm more than the single bond length (the van der Waals distance; section 7.2.1). The overall relationship of bond order (n) and bond length (r) can be described by the general equation (5.1)

$$\log_{10}(n)=(r_o-r)/c \qquad (5.1)$$

where r_o is the length of a standard single bond between two particular types of atoms and c is a constant depending on the atoms involved but not varying over a great range. For C–C bonds, we have $r_o=153.4$ pm, $c\approx72$ pm.

This equation has an intriguing and unexpected implication. The effect of either *doubling* or *halving* the order of a bond (e.g. 1 to 2, or 1 to 0.5) is rather uniform,

producing a change in length of 20 to 30 pm (half this in the radius); the change tends
to be rather smaller for first row elements, larger for later ones.

5.8 BOND LENGTH AND BOND ENERGY

The original assumption (section 2.3.3) that bond length is closely related to bond
strength can be directly studied in carbon compounds where reasonable values for
bond energies (strictly *bond dissociation energies*; section 2.3.1) can be calculated.
Within the bond order range 1.0–3.0, the graph (Fig. 5.26a) is precisely linear, but

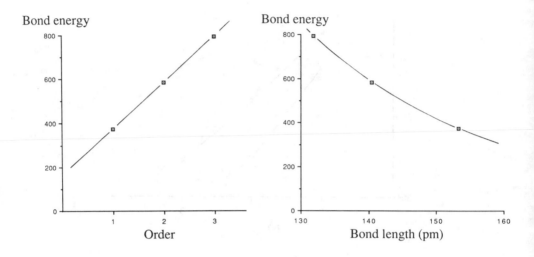

Fig. 5.26 — Correlation of bond order and (a) bond dissociation energy, (b) bond length.

the line does not pass through the origin. This shows that the two π-bonds contribute
about the same to the bond energy, but that the σ-bond is stronger, and therefore the
fall-off with increasing bond length (Fig. 5.26b) is more rapid. A detailed knowledge
of the effect of changes in bond length on energy is essential for molecular mechanics
calculations, but it has been found more effective to use empirical rather than
theoretical correlations (section 2.3.2).

A relationship would also be expected between bond length and an alternative
measure of bond strength — the force needed to stretch the bond or *stretching force
constant*. Indeed, a relationship has been proposed, Badger's rule:

> force constant$=a/$(length)^3+b
> (*a* and *b* are empirical constants).

The difficulty in applying this lies in obtaining the force constants (Box 10), and it is
doubtful that the available evidence, even for C-bonds, is accurate enough to prove
the rule and calculate the constants *a* and *b*. A recent study of C–H bonds has found a
straight-line relationship between the length and the vibration frequency v (Fig.

Box 10   Force constants

For a diatomic molecule, the relationship between vibration frequency (v in cm^{-1}) and force constant (k in $N\ m^{-1}$) is:

$$v = 1/2\pi c.\sqrt{k/\mu}$$

(c=speed of light, μ=reduced mass of the molecule)

and the force constant can be obtained directly from the vibration frequency. However, most vibrations of polyatomic molecules, involve more than one force constant. For example, the antisymmetric stretch in H_2O involves the atoms moving in the way shown

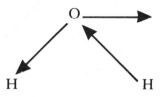

This is affected by H–O–H bending as well as H–O stretching. Determination of the fundamental information for a particular bond, the force constant, is therefore complicated and may be impossible. For the same reason, an i.r. frequency, which is easy to observe, generally does not relate precisely to one individual bond.

5.27). However, the range in bond length (caused mainly by the changes in hybridization discussed next) is tiny, 106–111 pm. The apparent straight-line relationship between r and v corresponds to a relationship between r and k^2, but the difference between k^2 and k^3 (as in Badger's rule) would be hardly detectable. Overall, the relationship between bond length and bond strength is well-established, but how they are quantitatively related is much less clear.

5.8.1   Covalent radius and hybridisation

Hybridization has a definite influence on bond length. The presence of an sp^2 carbon atom (forming a double bond) produces *single bonds*, mean C–C distance of 150.3 pm (X), shorter than the standard value of 153.4 pm. An sp atom has an even bigger effect (XI), giving a mean C–C distance of 146.6 pm (Allen *et al.* 1987). The best

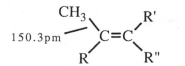

X (mean of 215 values)   XI (mean of 21 values)

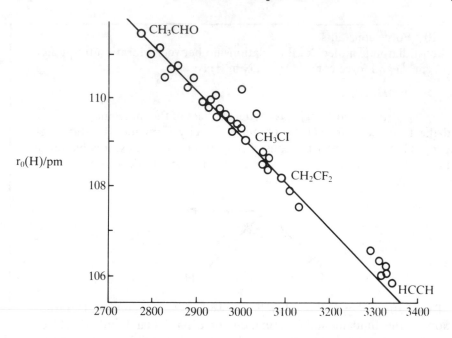

Fig. 5.27 — Relationship between C–H distance and νCH. Source: McKean, D. C. (1984) *J. Mol. Str.* **113** 251 (which identifies all the plotted points).

values for carbon radii, allowing for this are:

Hybridization	Radius (pm)
sp^3	76.7
sp^2 (also aromatic C)	73.6
sp	69.9

The difference arises because the s orbital has its maximum electron density nearer the nucleus than the p orbital. Thus, the more s character, the shorter the bond. Curiously with N, the shift is apparently in the reverse direction, giving an increase for a sp^2 N-atom from 70.2 to 72.6 pm.

5.8.2   Two problems
Our assumption of constant covalent radius is not entirely correct, as two series of measurements show.

(a) Fluoro-methanes have the following C–F distances

CH_3F	138.5
CH_2F_2	135.8
CHF_3	133.2
CF_4	132.3 pm

(A similar reduction occurs in C–Cl distances in chloro-methanes but only by about 2 rather than 6 pm.)

Clearly, standard radii cannot predict these distances precisely, though if we take the mean values, the deviations are not too severe. They are probably caused by the hybridization effect just discussed. A regular tetrahedral molecule, CX_4, has four identical bonds, each sp^3 hybridized. If the four substituents are not identical, and in particular if they differ very much in electronegativity, some bonds may have rather more s character and some rather more p character. The former should be rather shorter than the latter, though the effect should not be as great as when the hybridization changes.

(b) Bonds between electronegative atoms are generally longer than predicted. The distances in the following compounds can be compared to the predictions from standard radii:

Bond	Calculated	Observed	Compound
N–N	140.4	145.3	N_2H_4
N–O	136.9	146	H_2NOH
N–F	132.1	137.1	F_3
O–O	131.8	146.7	H_2O_2
O–F	127.8	141.3	F_2O
F–F	123.8	141.8	F_2

This substantial lengthening, as the electronegativity increases, correlates with an important effect on bond energy, making the O–O and F–F bonds weaker than would be expected. For the F–F bond, this weakness is easy to demonstrate, as its bond dissociation energy is actually less than for Cl_2. A complete explanation has not been given, but the simplest view is that the pairs of electrons on each atom repel each other. The discussion of the detailed bonding in O_2 and F_2 also throws light on this problem (section 5.9.1). F–O compounds in general contain very weak bonds, and this has curious structural consequences (section 9.5).

5.9 MULTI-CENTRE π-BONDING IN NON-METAL COMPOUNDS

The final section of this chapter deals with a group of compounds with more complicated π-bonding arrangements than those so far examined. These bonds usually extend over more than two atoms, i.e. are of the type known as *delocalized* (localized bonds being those between *two* atoms). They are also interesting because:
(a) The same bonding schemes apply to several compounds differing in the specific atoms involved, but with the same number of electrons. Thus, CN^- and N_2 both have 10 valence electrons and identical patterns of molecular orbitals. Such pairs of molecules are described as *isoelectronic* (i.e. containing different atomic species but having the same geometry and the same number of electrons). (b) Several of these compounds can exist with different numbers of electrons; not only NO_2^- (18 electrons) but NO_2 and NO_2^+ (17 and 16 electrons). With these compounds we can

make a direct study of the effect of changing the occupancy of the bonding and antibonding orbitals.

The compounds surveyed are listed in Table 5.1, grouped according to the number of atoms and of electrons.

Table 5.1 — Isoelectronic species

Species in brackets do not formally belong to the isoelectronic groups because they contain elements with different inner electron shells.

Diatomic	Triatomic	Tetratomic	Pentatomic
9 els.	16 els. Linear	24 els. Planar	32 els. Tetrahedral
$[N_2]^+$ $[CO]^+$	CO_2 $[NO_2]^+$ N_2O	$[NO_3]^-$ $[CO_3]^{2-}$	CF_4 $[BF_4]^-$ $[NF_4]^+$
CN	$[N_3]^-$ $[NCN]^{2-}$	$[BO_3]^{3-}$ COF_2 BF_3	F_3NO $([PO_4]^{3-})$
	$[NCO]^-$ $[N_2F]^+$	$[NOF_2]^+$ NO_2F (SO_3)	$([SO_4]^{2-}$ $[ClO_4]^-)$
	$([SCN]^-)$		
10 els.		26 els. Pyramidal	
N_2 CO $[CN]^-$			
11 els.	17 els. Bent	NF_3 $([SO_3]^{2-}$ $[ClO_3]^-)$	
$[O_2]^+$	NO_2		
12 els.	18 els. Bent		
O_2	$[NO_2]^-$ O_3		
13 els.	19 els. Bent		
$[O_2]^-$	$[O_3]^-$ NF_2		
14 els.	20 els. Bent		
$[O_2]^{2-}$ F_2	F_2O (Cl_2O)		

5.9.1 Diatomic molecules

The pattern of molecular orbitals in diatomic molecules like N_2 is very similar to that in C_2H_2 (section 5.5.1). These molecules are of course linear and the atomic orbitals are grouped into non-interacting sets:

Type	Contributing orbitals
σ	s, p_z (assuming z as molecular axis)
π_x	p_x
π_y	p_y

The σ orbitals on each N-atom can be hybridized sp. The inward-pointing orbitals produce bonding and antibonding σ-molecular orbitals, and the outward ones contain lone pairs; this is the only difference from C_2H_2, where these orbitals form bonds to the H-atoms. Each set of π orbitals overlap to give ψ_b and ψ_{ab} orbitals. The complete energy level diagram is shown in Fig. 5.28. N_2 contains 10 electrons, and clearly these occupy the five lowest orbitals: all the bonding and the two non-bonding (lone-pair) orbitals. Three pairs of electrons are in bonding molecular orbitals and none in antibonding ones, giving a bond order of 3, corresponding to the simple valence picture: N≡N. The bond is correspondingly short (109 pm) and strong (bond dissociation energy 945 kJ mol^{-1}).

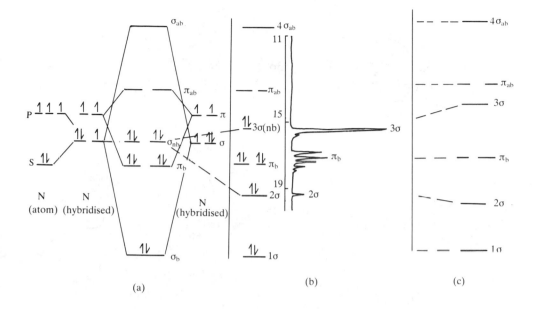

Fig. 5.28 — N₂: (a) simple molecular orbital diagram, (b) energy levels corrected for interaction of hybrid σ orbitals, (c) Photoelectron spectra and energy levels. Removal of an electron from the 1σ orbital requires too much energy to be observable, (d) energy levels in CO.

It is useful with N_2 to go beyond the approximation of treating the sp-hybrid σ orbitals as strictly localized, rather than interacting with each other. The most important effects of their interaction are found with the non-bonded orbitals. These are lowered in energy relative to the π orbitals, and also cease to be degenerate (of equal energy). The energies after interaction are shown in Fig. 5.28b, with their shapes in Fig. 5.29; the four σ orbitals are labelled 1σ to 4σ. The relative energies of the σ and π orbitals are difficult to establish from qualitative arguments, and energy diagrams often show the filled π orbital highest. Detailed calculations suggest that the $σ_{nb}$ lies just above the π, and this is confirmed by the photoelectron spectrum (Fig. 5.28c).†

These improvements in the molecular orbital description help us to understand the effect of ionizing N_2. The electron is removed from the highest occupied orbital (3σ). As this has some electron density between the N-atoms (Fig. 5.29), it is slightly bonding. Thus, ionization weakens the N–N bond slightly, giving N_2^+ a bond length of 112 pm.

The molecule CO has exactly the same orbital pattern as N_2, except that it is no

† Most textbooks use an alternative approach to the bonding in N_2, starting from separate overlap of 2s with 2s and $2p_z$ with $2p_z$; they sometimes then consider the result of mixing the s- and p-molecular orbitals, which produces the same final diagram as here (Fig. 5.28b). This alternative approach ignores the relationship between N_2 and C_2H_2 (for which sp-hybrids are generally used), and it conceals the lone-pair character of the 2σ and 3σ orbitals.

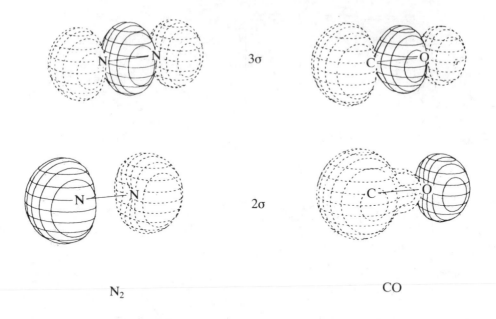

3σ

2σ

N₂   CO

Fig. 5.29 — Shapes of 'non-bonding' or lone-pair orbitals in N_2 and CO. Note for CO the
additional electron density near O in the 2σ orbital and near C in the 3σ orbital. Source: as Fig.
5.4 (Figs 15–16).

longer symmetrical. CO still has a bond order of 3, and the bond length (112.8 pm) is
only slightly longer than in N_2. The greater attraction of the O-atom for electrons
influences the energies of 2σ and 3σ. 2σ is lowered in energy and concentrated near
the O-atom, i.e. is predominantly its lone pair. 3σ is higher in energy and located
near the C-atom, though it is rather more diffuse than 2σ. This has two effects:

(a) 3σ becomes slightly antibonding, so that CO^+ is slightly shorter than CO (111.5
    pm).
(b) Both CO and N_2 can act as donors (Chapter 7), involving overlap between one of
    these lone pair orbitals and an empty (acceptor) orbital on another atom. When
    CO acts as a donor (Chapter 7), the electrons in 3σ rather than 2σ are
    preferentially donated, i.e. donation occurs from C not O; CO is also a better
    donor than N_2 as its 3σ orbital has a better overlap with the acceptor orbital.

   The O_2 molecule has the same orbital scheme as N_2 (Fig. 5.30a in the simple
approximation), but contains two more electrons. The lowest available empty
orbitals, are $\pi_{x, ab}$ and $\pi_{y, ab}$. To minimize electron repulsion, one electron is in each
orbital and, by Hund's first rule (Box 11), they are parallel. As they are in
antibonding orbitals, they reduce the total bond order from 3 (N_2) to 2 (O_2). This
corresponds to the simple bond picture, O=O, but the molecular orbital treatment
leads to a second and very important prediction: that O_2 should be paramagnetic,

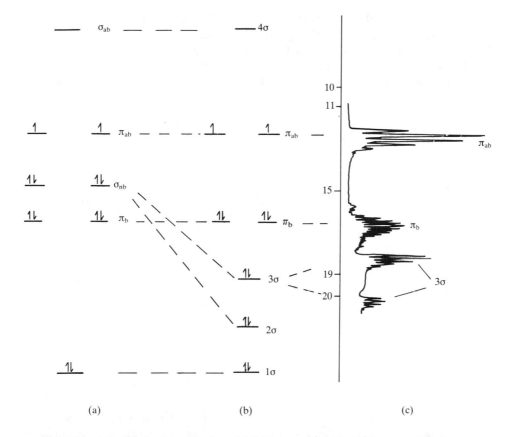

Fig. 5.30 — O_2: (a) simple molecular orbital diagram, (b) corrected for attraction of non-bonded electrons by oxygen atoms, (c) Photoelectron spectrum of O_2. Both levels marked 3σ correspond to removal of an electron from 3σ, depending on whether the remaining electron is parallel or anti-parallel to the two electrons in π_{ab}.

Box 11   Hund's rules

The most stable configurations of electrons in atomic orbitals accords with Hund's three rules.

1. Whenever possible, electrons occupy separate orbitals rather than being paired up. This reduces electron–electron repulsion because the electrons are separated as far as possible from each other. The principal exceptions to this rule are the 'low-spin' complexes of transition metals (section 8.2.2).

2. The electrons are aligned parallel rather than anti-parallel. This increases the 'electron correlation' (section 6.5).

3. The electrons occupy the orbitals of maximum orbital angular momentum, subject to the constraints of rules 1 and 2.

because of its unpaired electrons (Box 5). This is fully confirmed by experiment. A less important effect in O_2 is that the greater electronegativity of oxygen compared to nitrogen causes both 2σ and 3σ to lie below the π_b levels (as happens also in F_2) (Fig. 5.30b). This is confirmed by the photoelectron spectrum (Fig. 5.30c).

A remarkable range of O_2 species can be obtained (shown in Fig. 5.31), and their

		O_2^+ oxygenyl ion	O_2 oxygen	O_2^- superoxide ion	O_2^{2-} peroxide ion
π-orbital occupancy	π_{ab}	↑ _	↑ ↑	↑↓ ↑	↑↓ ↑↓
	π_b	↑↓ ↑↓	↑↓ ↑↓	↑↓ ↑↓	↑↓ ↑↓
total bond order		2.5	2.0	1.5	1.0
bond length (pm)		112.3	120.7	128	149
unpaired electrons		1	2	1	0
paramagnetic?		Yes	Yes	Yes	No

Fig. 5.31 — Bond order, bond length, numbers of unpaired electrons and paramagnetism for O_2 and related species.

properties exactly reflect the predictions based on the successive addition of electrons to the π_{ab} orbitals. In particular, the bond lengths mirror the bond orders. With the final member of the series, O_2^{2-}, the antibonding π orbitals are full, and the π-contribution to the bond order is zero. Furthermore, the average of an antibonding and a bonding orbital is slightly antibonding, because of the effect of overlap (Box 7), and the mutual repulsion of the electrons in each orbital. This gives one explanation for the weak and long bonds between highly electronegative elements. Fluorine, F_2, is isoelectronic with the peroxide ion, O_2^{2-}, but its bond is shorter because the fluorine atoms have larger nuclear charges.

5.9.2 Triatomic molecules — CO_2

By far the largest group of isoelectronic species (Table 5.1) are the linear molecules with 16 electrons, of which CO_2 is typical. Because it is linear, the orbital sets are exactly as for N_2. The σ-symmetry orbitals are sp hybridized and overlap to give two pairs of $\sigma_b + \sigma_{ab}$ orbitals, leaving two lone pair orbitals (σ_{nb}) (Fig. 5.32). The four σ_b

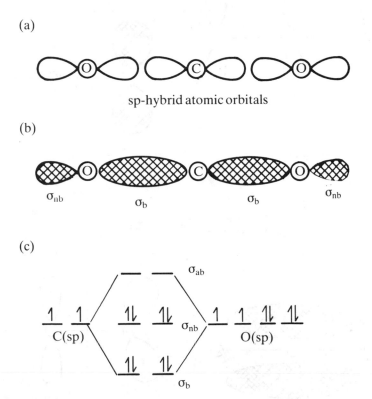

(a)

sp-hybrid atomic orbitals

(b)

(c)

Fig. 5.32 — σ orbitals in CO_2 (a) atomic orbitals, (b) molecular orbitals, (c) energy levels and occupancy.

and σ_{nb} orbitals are full (four pairs of electrons). The two equivalent sets of π atomic orbitals each comprise three orbitals, one per atom (Fig. 5.33a). Each must therefore give *three* molecular orbitals. Predicting their energies and shapes is not automatic, but in this case the most obvious arrangement is correct: one orbital of low energy (π_b), one intermediate (π_{nb}), and one high (π_{ab}) (Fig. 5.33b).† In the most favourable combination, the three atomic orbitals are added in phase, while the highest energy orbital is of antibonding character between each pair of atoms. The non-bonding orbital could be described in two ways involving one in-phase and one out-of-phase overlap ($\psi_1 + \psi_2 - \psi_3$ or $\psi_1 - \psi_2 + \psi_3$), but the form that corresponds to the molecular symmetry is the sum of these ($\psi_1 - \psi_3$); the orbitals ψ_1 and ψ_3 are too far apart for their overlap to produce appreciable antibonding (Fig. 5.33c–f). It is worth noting

† Contrast the orbital energies for an equilateral triangle of atoms (section 9.7.4).

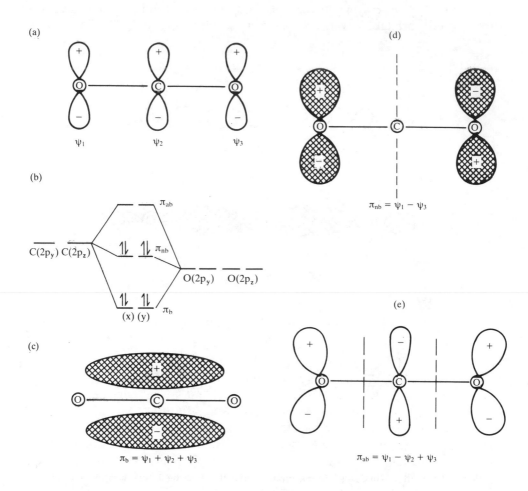

Fig. 5.33 — π orbitals in CO_2 (a) atomic orbitals (one of two equivalent sets), (b) energy levels, (c)–(e) molecular orbitals π_b, π_{nb}, π_{ab}.

that the π orbitals have successively 0, 1, and 2 nodal planes perpendicular to the molecular axis.

With 8 electrons in the σ orbitals, the remaining 8 are equally divided between π_x and π_y. Thus each π_b and π_{nb} is occupied (Fig. 5.34). As each set of orbitals covers three atoms and contains four electrons, they are described as *three-centre/four-electron* molecular orbitals. In total, 8 electrons are bonding (two σ- and two π-pairs). To calculate the overall bond order, we must allow for the number of bonds. Thus, in CO_2 the order of each C–O bond is 2.0.

5.9.3 Seventeen and eighteen electrons: NO_2; NO_2^-; O_3

NO_2^+ (XII), one of the ions isoelectronic with CO_2, is particularly interesting because it is possible to add one and then a second electron, giving NO_2 and NO_2^-

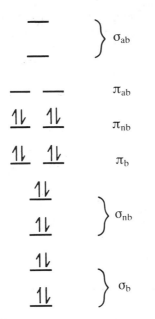

Fig. 5.34 — CO_2: overall energy levels and occupancy.

respectively. Clearly, the first added electron goes into the lowest available empty orbital: one of the π_{ab} orbitals. We might expect the only effect to be a reduction in bond order from 2 (8 bonding electrons for 2 bonds) to 1.75 (7 net bonding electrons). But, because NO_2 is triatomic, it can bend. The effect of bending is to make one π_{ab} orbital less unstable than the other.† This is clearly an advantage, because the single antibonding electron can occupy the more stable orbital. As a result, the NO_2 molecule (XIII) has a bond angle of 134.3° and N–O distance of 119.7 pm, compared to 115 pm in NO_2^+ (very similar to the C–O distance of 116.3 pm in CO_2). NO_2 is also, of course, paramagnetic. Adding another electron gives NO_2^- (XIV). This added electron pairs up with the odd electron of NO_2, giving NO_2^- 6 net bonding electrons, a bond order of 1.5, O–N–O angle of 115.4° and N–O distance of 123.6 pm.

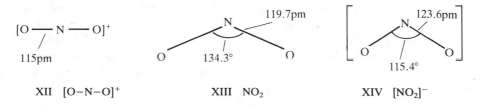

XII $[O–N–O]^+$ XIII NO_2 XIV $[NO_2]^-$

Once a triatomic molecule bends, the separation of σ, π_x and π_y is no longer valid.

† This is an example of a Jahn-Teller distortion (section 8.2.4).

One set of orbitals (say p_x) is independent, but s, p_y, and p_z all interact. Thus, it is better to treat NO_2 and especially NO_2^- as sp^2 hybridized. In NO_2^-, the σ-molecular orbitals (Fig. 5.35a) include two bonding orbitals (and corresponding $σ_{ab}$ orbitals),

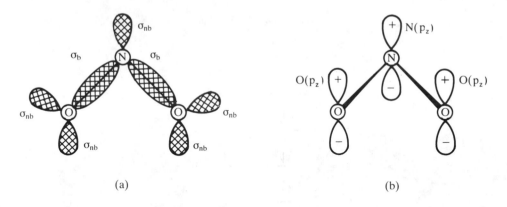

(a)   (b)

Fig. 5.35 — Molecular orbitals in NO_2^- (a) occupied σ orbitals (b) π-atomic orbitals.

and five lone pair ($σ_{nb}$ orbitals), occupied by 14 electrons. The three p_z orbitals (Fig. 5.35b) combine to give the same set of π-molecular orbitals as in CO_2, and the remaining four electrons occupy the $π_b$ and $π_{nb}$ orbitals as before. The $π_b$ pair contribute 0.5 to the bond order for each N—O bond, giving each bond a total order of 1.5.

Considering NO_2 in the same way as NO_2^-, the odd electron occupies the $σ_{nb}$ orbital on N. It is easy to see how the repulsion between this one electron and the two pairs in $σ_b$ orbitals is smaller than for the pair of electrons in this orbital in NO_2^-. This accounts for the 134.3° angle in NO_2, but fails to explain why the absence of one electron from the apparently non-bonding ($σ_{nb}$) orbital increases the bond order to 1.75. A detailed analysis shows that increasing the angle changes the precise energy levels of the σ-molecular orbitals, leading to this orbital being partly antibonding in character (which is easy to see when NO_2 is considered as derived from NO_2^+).

O_3 is another 18 electron species, isoelectronic with NO_2^- (with bond angle 116.8° and O—O distance 127.8 pm). It can accept a further electron into the $π_{ab}$ orbital. This gives O_3^-, the ozonide ion, with 19 electrons and a bond order of 1.25. Crystals containing this ion are unstable, but the structure of the ion has recently been determined; it has O—O of 128.6 pm, O—O—O of 119.5°, moving further in the direction of weaker bonds (Hesse & Jansen 1988).

With one more electron, F_2O has all its bonding π-electrons counterbalanced by antibonding electrons. As already noted (section 5.9.1), this leads to net antibonding and provides an explanation for its exceptionally long bonds (F—O 140.5 pm, F—O—F 103°).

5.9.4   Tetratomic species – NO_3^-

NO_3^- is a planar molecules with 24 electrons. As with NO_2^-, only one set of p orbitals is of π-symmetry, and the σ orbitals can form sp_2 hybrids. They produce three $σ_b$

orbitals, and each O-atom has two σ_{nb} orbitals; these contain a total of 18 electrons (Fig. 5.36a).

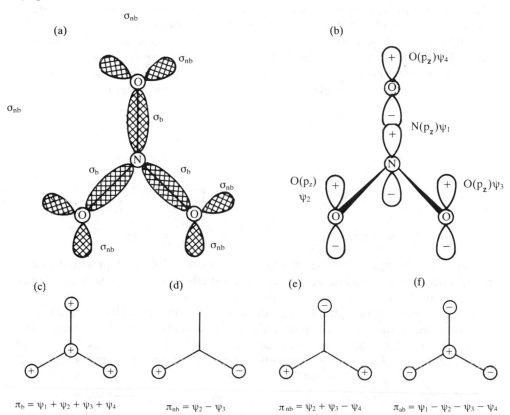

Fig. 5.36 — Orbitals in NO_3^- (a) σ-molecular orbitals, (b) π-atomic orbitals, (c–f) π-molecular orbitals, showing the sign combinations for each orbital, looking down on the NO_3^- plane.

The four π-atomic orbitals (Fig. 5.36b) give four molecular orbitals, delocalized over the molecule. It is easy to see that the most stable molecular orbital will be that shown in Fig. 5.36c

$$\psi_b = \psi_1 + \psi_2 + \psi_3 + \psi_4$$

and that the highest energy one is that in Fig. 5.36f

$$\psi_{ab} = \psi_1 - (\psi_2 + \psi_3 + \psi_4).$$

Working out the form and energies of the two others requires consideration of the molecular symmetry, beyond the scope of this book (Loewe & MacArthur 1977).

This shows that they are in fact non-bonding orbitals of equal energy. They can be represented in various ways, of which two are shown in Fig. 5.36d–e. The final six electrons occupy π_b and the two π_{nb} orbitals. As the one bonding pair is delocalized over three bonds, it contributes 0.33 to the total bond order of 1.33.

Of the other 24 electron species (Table 5.1), CO_3^{2-} has the same orbital scheme as NO_3^-, but the bonding in BO_3^{3-} and BF_3 raises an interesting question. Both clearly have 24 electrons and are planar, like NO_3^-, but on a simple bond diagram would be regarded as containing single bonds (XV, XVI). These diagrams conceal the

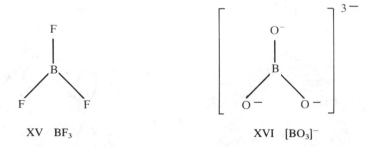

XV   BF₃   XVI   [BO₃]⁻

presence of the empty p_z orbital on B in each molecule. This overlaps with the filled p_z orbitals on F or O. The difference between BF_3 and NO_3^- lies in the lesser attraction of B than N for electrons. If this attraction were nonexistent, the π_b orbital would be of the same energy as the π_{nb}; the three non-bonding orbitals would just be combinations of the three lone-pair orbitals of π-symmetry, and the π-bond order would be zero. If the attraction were as strong as in NO_2^-, BF_3 would also have a total bond order of 1.3. Fig. 5.37 shows the effect of these alternatives on the energy level

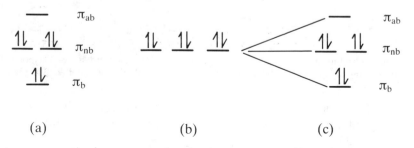

(a)   (b)   (c)

Fig. 5.37 — π-molecular orbitals in (a) NO_3^-, (b) BF_3 with no involvement of the p_z on B, (c) BF_3 including interaction between this p_z and the p_z orbitals on the fluorine atoms.

diagram. Comparison of the bond length in BF_3 (130 pm) with the predicted value from standard radii (145 pm) suggests that it has considerable double bond character, and is best treated as an analogue of NO_3^-. (It is further discussed in section 7.3.3.3.)

Adding two electrons to NO_3^- destroys all the π-bond stabilization. We would expect the molecule to become non-planar, to reduce the overlap and repulsion between the π-orbitals. This is confirmed in NF_3, the only example of a 26 electron molecule containing just first row elements. The ions SO_3^{2-} and ClO_3^- also have 26 valence electrons, but the empty d-orbitals of the second row central atoms have an important influence on the bonding (section 9.4).

5.9.5 Pentatomic molecules — CF$_4$

All the atomic orbitals in a tetrahedral molecule interact together. In the simple description we are using, each atom can be regarded as sp^3 hybridized. All the electrons are in σ orbitals, eight in the σ-bonds and eight on each atom as lone pairs, and the bond order is 1.0. The bonding is identical to that in CH$_4$, apart from the lone pairs. Parallel second-row compounds exist, such as ClO$_4^-$, but again their bonding is complicated by the d-orbitals on chlorine.

FURTHER READING

Molecular orbital treatment of bonding.

Purcell, K. F. & Kotz, J. C. (1977) *Inorganic chemistry*. Saunders, Philadelphia. (Particularly thorough, though making no concessions to the inexperienced reader.)

Kettle, S. F. A. (1986) *Symmetry and Structure*. Wiley, Chichester. (Provides a detailed analysis of the application of symmetry to covalent bonding).

Jorgensen, W. L. & Salem, L. (1973) *The organic chemist's book of orbitals*. Academic Press, New York.

Structure, geometry and dimensions

Wells, A. F. (1984) *Structural inorganic chemistry*. Clarendon Press, Oxford, (5th ed). (Gives some discussion of covalent molecules and VSEPR. The latter is also described in many inorganic textbooks.)

For tabulation of bond distances in covalent molecules, see Allen, F. H., Kennard, O., Watson, D. G., Brammer, L., Orpen, A. G., & Taylor, R. (1987) *J. Chem. Soc. Perkin II* p. S1.

Bassindale, A. (1984) *The third dimension in organic chemistry*. Wiley, Chichester.

Testa, B. (1971) *Principles of organic stereochemistry*. Marcel Dekker, New York, (Discuss conformational preferences and torsion angles.)

π-bonded molecules

Purcell & Kotz are again particularly informative, though their approach differs in detail from that adopted here.

Text references

Hesse, W. & Jansen, M. (1988) *Angew. Chem. Int., Ed. Engl.* **27**, 1341.
Loew, L. M. & MacArthur, W. R. (1977) *J. Amer. Chem. Soc.* **99** 1019.

Part III
Real bonds and real compounds

6

Metals

6.1 INTRODUCTION

In studying the chapters of Part II, readers might understandably think that the bonding models were excellent, as long as they were restricted to sodium, sodium chloride, or methane — the corners of the bond triangle. In Part III, our survey is extended to the whole range of chemical compounds, including especially those with intermediate bonding, or which combine more than one bond type. The bonding principles described in Part II remain valid. Within their framework, the important differences are produced by the changing electronegativity of the components of the A_mB_n system. Some of these compounds are particularly interesting both in their structures and properties, and these are the special concern of Part III.

6.2 DETAILED BAND STRUCTURES OF METALS

Comparison of Chapter 5 (covalent compounds) with Chapter 3 (simple metals) may have led some readers to notice a modest contradiction. In covalent bonding, one of the key principles is that molecular orbitals cannot be created or destroyed, only combined and reorganized. However, in considering metallic bonding in sodium, its 3s orbitals were simply ignored. This must be wrong. Consider the gradual coalescence of sodium atoms with one electron in each 3s orbital, into sodium metal with the same electrons in delocalized orbitals (Fig. 6.1). Clearly the delocalized metal orbitals must in reality be molecular orbitals formed from combinations of the 3s atomic orbitals. This chapter re-examines the band structure of metals, using this viewpoint, and also considers a less obvious problem caused by the repetitive arrangement of the metal atoms in the crystal.

Deriving the delocalized metal orbitals of sodium from combinations of the 3s atomic orbitals poses no problems, as we do not need to determine the shape of the individual orbitals. In a crystal containing N sodium atoms, a band of N molecular orbitals will be produced from the 3s atomic orbitals. Its N valence electrons will half-

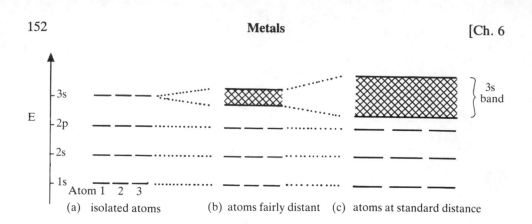

Fig. 6.1 — Formation of 3s band of orbitals in Na metal (a) energy levels of isolated atoms, (b) fairly distant atoms, (c) atoms at standard metallic distance.

fill this band (Fig. 6.2a), and the deductions made in Chapter 3 about the properties of a compound with such an electronic structure are still valid. However, turning to Mg metal reveals a problem. Its $2N$ electrons should completely fill the 3s band of molecular orbitals, leaving no empty orbitals just above E_{max} for electron promotion during electrical conduction, optical absorption, etc. The difficulty is resolved if we consider also the 3p orbitals of Mg. These also combine to form a band of molecular orbitals. Because both the 3s and 3p orbitals are fairly diffuse, orbitals on adjacent atoms overlap well. Both bands of molecular orbitals have a considerable range in energy, and overlap each other (Fig. 6.2b). Even though there are enough electrons to fill the 3s band, vacant orbitals remain available to provide metallic properties.

First row transition metal atoms give rise to similar bands of orbitals, derived from the 3d as well as the 3s and 3p atomic orbitals. However, the bands formed from

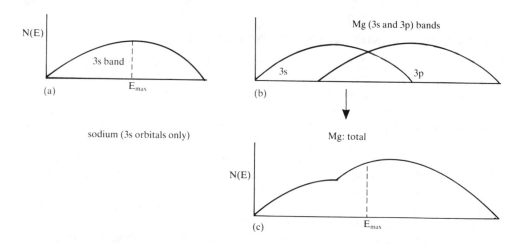

Fig. 6.2 — Occupancy of molecular orbitals in metals (a) half-filled 3s band in sodium (b) 3s and 3p bands in Mg (shown distinct) (c) overall band structure in Mg, showing E_{max}. Note that Na will also have a 3p band, contributing to the available orbitals at energies rather above E_{max}.

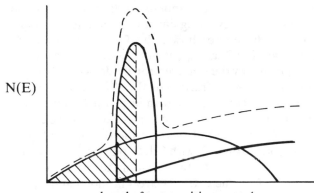

bands for transition metal

Fig. 6.3 — Bands in a transition metal. The dashed line shows the total number of available energy levels.

d orbitals are somewhat different from the s- and p-bands. The d-electrons are strongly held by the individual metal atoms. As a result, their overlap with orbitals on other atoms is more limited. The d-band is therefore more concentrated (Fig. 6.3). Note that here, as for Mg, the *total* number of available orbitals is the sum of those in the individual bands.

The influence of the crystal structure on the energy levels in a metal is more subtle. Electrons with a particular energy in the delocalized orbitals are moving at a corresponding speed (and therefore can be viewed as waves with a particular wavelength). At certain speeds and directions, they interact with the lattice, are diffracted, and undergo interference in exactly the same way that X-rays are diffracted by a repeating lattice (see Chapter 2 and Fig. 2.3). As a result, these electron energies/directions are prohibited. The effect on $N(E)$ in a simple metal is seen in Fig. 6.4. The small peak occurs at the first energy for which diffraction occurs,

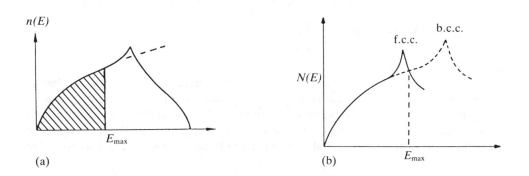

Fig. 6.4 — The modification of $N(E)$ by the diffraction of electrons in the crystal lattice. (a) For Na. The dashed line shows the original parabola, and the shaded area corresponds to E_{max} for one electron per atom. (b) Comparison of $N(E)$ for c.c.p. and b.c.c. structures. At the value of E_{max} shown by the dotted line, the b.c.c. structure becomes the more stable.

and the rapid reduction in $N(E)$ corresponds to its increasing importance as the energy increases. Higher energies are again allowed, and further bands appear (not shown in Fig. 6.4). As the figure shows, the free-electron values of $N(E)$ are unaffected at low energy. For Na, E_{max} is at 0.88 of the peak value in the b.c.c. structure, and this explains why the free-electron model works very well for sodium.

The precise effect of electron diffraction on the $N(E)$ curve depends on the crystal structure. As examples, the curves are shown in Fig. 6.4b for the c.c.p. and b.c.c. structures. Because the envelopes of the two bands intersect, we expect a correlation between the number of metallic electrons and the structure adopted. This is well illustrated by the effect of alloying Au with Cd. In pure gold, each Au atom provides one metallic electron in a c.c.p. structure, leading to the picture of Fig. 6.4a. In replacing an Au atom by Cd, one extra electron is added and E_{max} rises. At 1.36 electrons per atom, the band is filled to the point marked by the dotted line, and the b.c.c. structure becomes more stable, as the electrons have a lower total energy in this structure. This corresponds to 36% Cd in Au (1 electron from each atom + 0.36 extra electrons from Cd). Most satisfactorily, the stability range of the simple Au/Cd alloy in which Cd just replaces atoms of Au is 0–35% Cd, and above this a structural change to b.c.c. takes place.

For transition metals the precise band structure is much more complicated than Fig. 6.3 suggests. Indeed, even for Au/Cd, the replacement of a +1 ion by a +2 ion must affect the shape of the band. The results of detailed calculations for transition metals in each of the three main metal structures are shown in Fig. 6.5a–c. As the number of electrons increases, the relative stability of the three structures changes (Fig. 6.5d). This leads to the prediction of a structure sequence

$$hcp \rightarrow bcc \rightarrow hcp \rightarrow ccp$$

The observed room-temperature structures of these metals are shown in Fig. 6.6. They agree most satisfyingly with these predictions.

6.3   METAL PROPERTIES

6.3.1   Cohesive energy

The cohesive energy (or sublimation energy) of a metal is the energy needed to disrupt the metallic crystal, converting it into gaseous atoms. It is one of the most useful properties in understanding its chemical and physical behaviour. With high values of cohesive energy come high melting and boiling points, and high strength, all related to the strong metal-metal bonding.† In chemical reactions, a low cohesive

---

† Sublimation enthalpy and boiling point are directly linked through Trouton's constant $(\frac{\Delta H}{T})$, theoretically 85 J K^{-1} mol^{-1} (the entropy for one mole of gaseous atoms). Values for metals generally come close to this. Although melting point and Young's modulus are not so directly linked to the sublimation energy, they follow the same trend.

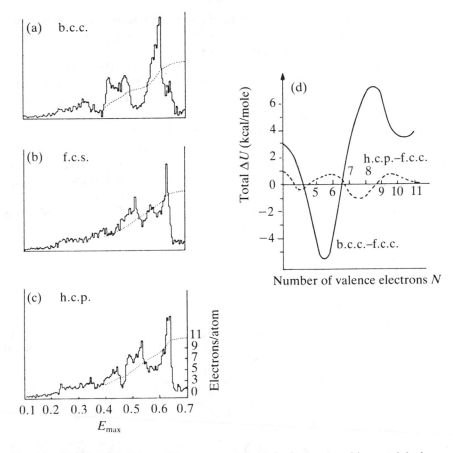

Fig. 6.5 — Detailed calculations of electron energy levels for first row transition metals in three different structures (a–c) individual energy levels (solid) and $N(E)$ (dashed line), (d) Predicted relative stability of the three structures. Reproduced with permission from Pettifor, D. G. (1977) *Calphad*, **1**, 305.

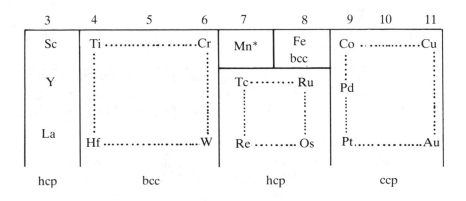

Fig. 6.6 — Structures of transition metals. * Mn has a complex structure.

energy, as for the alkali metals, leads to high heats of formation for compounds, as appears in calculations using Born–Haber and related cycles. The comparison of AuF and KF (Sect 5.6.3) illustrates this clearly.

As the variation between the properties of different metals suggests, cohesive energies are extremely variable, from 61 kJ mol^{-1} for Hg to 850 kJ mol^{-1} for W. Some systematic trends are apparent. Thus, for the Group I and Group II metals, values decrease down the groups, while for the transition metals they increase and then decrease again across each period, though with substantial irregularities (Fig. 6.7). The energies also increase from the first to the second and third rows. Conventionally, this behaviour in the transition metals has been associated with the addition of electrons, first to the lower energy part of the d-band leading to stabilization and additional bonding, and then to the higher energy part of the band cancelling the previous stabilising effect (e.g. Cox 1987, p.70). The irregularities in

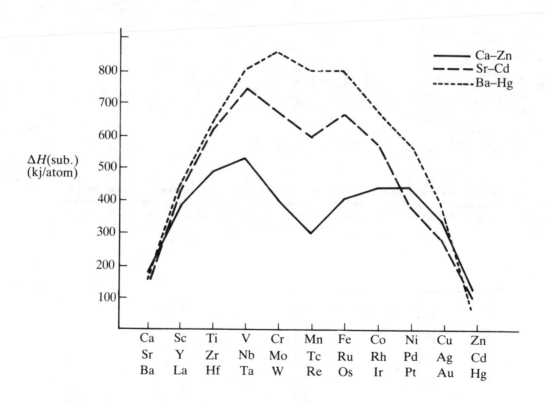

Fig. 6.7 — Sublimation energies for the three transition metal series. Reproduced with permission from Cox (1987), p. 69.

the centre are qualitatively associated with interactions between the unpaired electrons. Calculations of sublimation enthalpies have been moderately successful for the alkali metals, but are not yet able to reproduce the values for the transition metals.

However, we can also consider the cohesive energy of a metal in a different way (Alcock 1990). In reality, the sublimation energy is less directly linked to the metallic bonding itself than its influence on the chemical and physical properties of metals might suggest. Consider the cycle shown in Fig. 6.8 for the sublimation of a metal

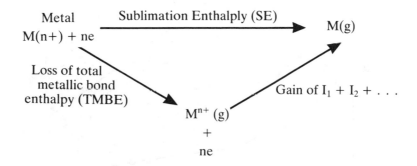

Fig. 6.8 — Thermochemical cycle for the sublimation of a metal (existing in the metal as M^{n+}).

which loses n electrons to its delocalised metallic energy levels. The crucial energy for metallic bonding is the energy (or the enthalpy) with which these electrons hold the metal ions together. This enthalpy can be called the *total metallic bond enthalpy* (TMBE), and is the enthalpy change when the solid metal is converted into gaseous metal *ions* and *electrons*, rather than gaseous atoms. As Fig. 6.8 shows, this is related to the standard sublimation enthalpy (SE) by the equation:

$$TMBE = SE - I_1 - I_2 - .. - I_n \tag{6.1}$$

The metallic properties with which the TMBE should be most closely related are the charges on the metal ions, and their spacing, controlling the positive potential experienced by the delocalized electrons. This is immediately apparent in Fig. 6.9 which shows the linear plot of TMBE with metallic radius for the alkali metals. As it also shows, the sublimation enthalpies are broadly linked to the metallic radius, but they are affected by variations in the ionization potential arising from the electronic structure of the metal atoms. The precise correlation of metallic bonding energy and metal density that this plot implies gives experimental support to the assumptions used in section 3.2.2 to explain the relative stability of close-packed and b.c.c. metal structures.

6.3.1.1 Sublimation energies for lanthanide elements

In the lanthanide elements (La–Lu), the seven 4f orbitals fill with electrons. As these are inner orbitals, their chemical influence is weak, and the lanthanides are

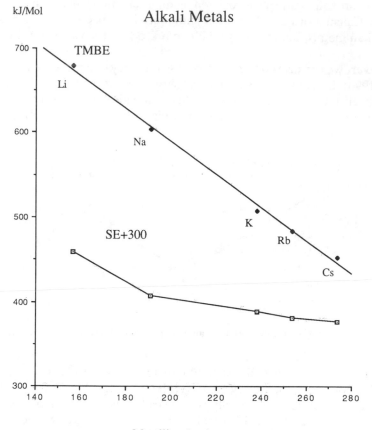

Fig. 6.9 — Sublimation enthalpies and total metallic bond energy (TMBE) for the alkali metals, plotted against metallic radius (actual values in their b.c.c. structures rather than the 12-coordinate values of Appendix A.3).

extremely similar in their general chemistry. In particular, they all form stable M^{3+} ions, and their radii (metallic and ionic) decrease smoothly from La to Lu. Despite this, some of their basic thermodynamic properties show surprising irregularities (Fig. 6.10). The sublimation enthalpy in particular varies between 470 (Ce) and 150 kJ mol^{-1} (Yb). However, the TMBE is still a linear function of radius (Fig. 6.11), though the 3+ charge of the ions gives the line a steeper slope and much larger intercept than for the alkali metals.

The obvious discrepancies for Eu and Yb are of particular interest in relation to their metallic bonding. These two elements have the largest third ionization potentials (Fig. 6.10). As a consequence, they are most easily reduced from M^{3+} to M^{2+} in their compounds. Considerable experimental evidence also shows that as metals, they exist as M^{2+}, i.e. I_3 is too large for the release of a third electron to the delocalized metallic orbitals to be compensated for by the gain in metallic bonding

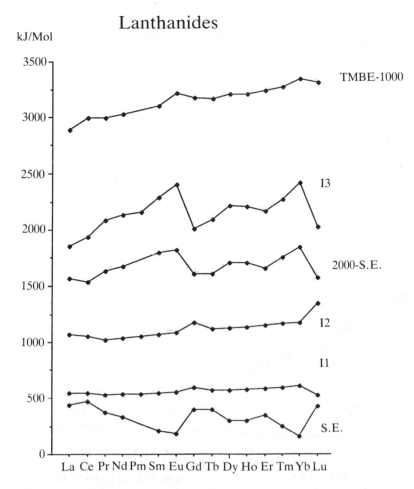

Fig. 6.10—Thermochemical properties for the lanthanides; Based on Johnson, D. A. (1969) *J. Chem. Soc. (A)*, p. 1525, with figures for Pm and Lu from sources quoted there and in Johnson (1977).

energy. Thus, they lie off the straight line in Fig. 6.11, but if their TMBE is recalculated for them as M^{2+} ions, they lie on the same straight line as do the Group II metals. Eu and Yb have b.c.c. and c.c.p. structures respectively, in contrast to the h.c.p. structures of most of the remaining lanthanides (at room temperature); their metallic radii are also abnormally large. Yb is the closer to behaving like a normal lanthanide, and it shows very unusual behaviour under a pressure of 40 kbar. It undergoes a structural change to b.c.c., with a 3% decrease in volume. The metallic radius decreases to 175 pm, bringing it into line with the other lanthanides. The ions in the metal have become M^{3+}!

The relationship in equation (6.1) and Fig. 6.8, coupled with the smooth change in TMBE, implies that irregularities in sublimation enthalpy arise as the inverse of those in the ionization potentials. This can be seen in Fig. 6.10, by comparing I_3 and

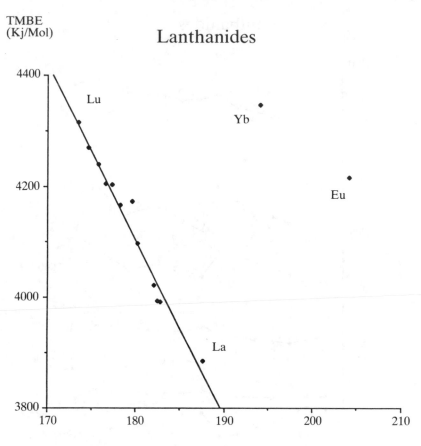

Fig. 6.11 — TMBE plotted against 12-coordinate metal radius for lanthanides.

−SE. The principal feature of I_3 is its high values for Eu and Yb, whose effect on metallic properties has just been noted. This arises because their M^{2+} ions contain either f^7 or f^{14} configurations, and *electron correlation* causes the removal of a further electron to be particularly difficult. This is important in relation to the chemical behaviour of the lanthanides, though it is not directly related to metallic behaviour. It is therefore discussed in the Appendix to this chapter.

6.3.1.2   *Sublimation energies for transition metals*

Thermochemical data for the first row transition metals are shown in Fig. 6.12; Fig. 6.13 includes TMBE plots for them as M^{3+} ions. These plots are linear for the first part of the series (Sr–Fe), showing that the ions in the metals are indeed M^{3+} as far as their metallic bonding is concerned. The later values fall above this line because the metals begin to behave as M^{2+} ions.

These curves allow us to understand the irregular form of the sublimation energy plots (Fig. 6.7) with which this section started. Following equation (6.1), the

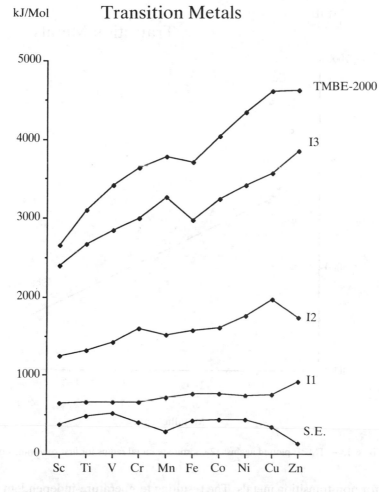

kJ/Mol   Transition Metals

Fig. 6.12 — Thermochemical data for first row transition metals.

sublimation energies are jointly affected by the TMBE and the ionization potentials. The rising parts of the curves result from the decreasing size of the metals, and their increasing ionization potentials. The decreases at the end appear because the very high I_3 values prevent the metal ions from becoming M^{3+} (as also with Ca, Sr, Ba). The irregularities in the middle reflect the steps in I_3, seen in Fig. 6.12 (for the first row elements). Just as with the lanthanides, these steps arise from the effects of electron correlation (see Appendix). The virtual disappearance of the step for the third row elements suggests that the influence of electron correlation on I_3 must be much smaller than for the other rows. However, this cannot be tested experimentally, as I_3 values are not known for the elements Hf-Pt.

6.3.2   Paramagnetism and ferromagnetism
The part-filled d-bands in the transition metals have one specific effect on their properties. The number of energy levels at E_{max} ($N(E_{max})$) is considerably larger

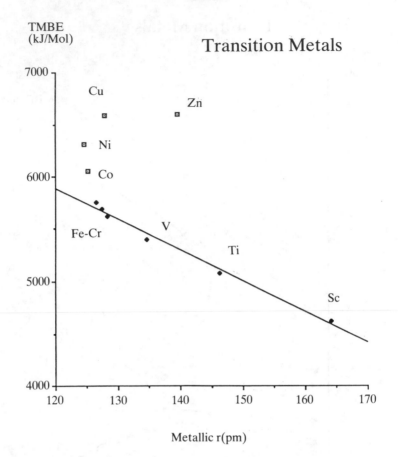

Fig. 6.13 — TMBE plotted against 12-coordinate metal radius for first row transition metals.

than for non-transition metals. The resulting temperature-independent paramagnetism is therefore much greater (Fig. 6.14; cf. Fig. 3.11). The same tendency for electron spins to be aligned, that causes the steps in I_3 and the irregularities in sublimation energies, causes Fe, Gd, and a few other metals to show *ferro*magnetism. Their electrons align themselves spontaneously rather than under the influence of a magnetic field, because of the favourable correlation energy for this process. The energy bands for 'up' and 'down' electrons are then as shown in Fig. 6.14, but arise spontaneously rather than under the influence of a magnetic field. In addition, the overlap of the d orbitals on different atoms allows this alignment to extend from one atom to another. The two metals for which this property is most marked, Fe and Gd, both have half-filled shells in their 3+ state. Alignment of the spins on different atoms therefore gives the maximum possible correlation energy. The absence of electrons with the opposite spin means that any electron which spontaneously changes direction finds no parallel electrons whose correlation energy it would enhance.

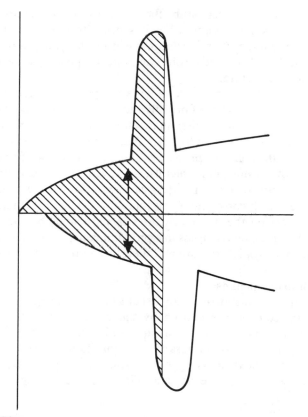

Fig. 6.14 — Paramagnetism in a transition metal. The large $N(E)$ causes strong paramagnetism.
Compare Fig. 3.11.

6.4 METAL ALLOYS

Judged simply by the number of element combinations, alloys of two metals outnumber all other binary compounds, though they are virtually ignored by chemists. Their properties reflect the subtle influence of the changing number of electrons in the delocalized metal orbitals. This leads to the appearance of such properties as ferromagnetism and superconductivity, as well as the mechanical strength with which the metallurgist is often concerned. For chemists, the remarkable variety of structure is probably their most intriguing feature.

6.4.1 Solid solution

The simplest type of alloy structure is the *solid solution*. It is also the most distinctive, virtually without counterpart in other chemical compounds. Solid solutions consist of varying proportions of two (sometimes more) metals, prepared by fusing the constituents and cooling the melt. They have the structure of the main component, with some of its atoms randomly substituted by atoms of the other component, exactly as for a solid dissolved in a liquid. Because the metallic bonding relies on the

attraction of delocalized electrons by the positively charged metal ions, the replacement of one metal ion by another often has little effect on the bonding. Indeed, not
merely are solid solutions found of one metal in another, but many metal alloy phases
have a range of stability caused by one component replacing some of the atoms of the
other in the alloy structure.

For solid solutions to occur over a wide composition range, two conditions are
necessary: that the metals are of broadly the same type, e.g. both transition metals,
and that their sizes are not too dissimilar. Thus, Cu and Au form solid solutions over
the entire range from 100% Cu to 100% Au. The formation of solid solutions often
has a systematic effect on alloy properties. The electronic specific heat, the maximum
magnetization of ferromagnetic metals, and even the critical temperature (T_c) of
superconducting metals (Box 12) may vary systematically with composition (Fig.
6.15). However, such correlations are found only when the changing composition
has a small effect on the band structure, in particular the d-electron band. The
presence of transition metal atoms in a predominantly s-p metal causes complicated
changes in the electronic structure and therefore in the properties of the alloy.

6.4.2 Electron compounds
Among the variety of distinctive alloy structures, two principal groups appear. In the
first, structural control is exercised by the number of metallic electrons. They
normally contain a post-transition metal (e.g. Zn, Sn, Pb, but also Be, Al) alloyed
with another of the same sort or, less often, a transition metal or pre-transition metal.
The *brasses*, the alloys of copper and zinc, are typical examples of *electron
compounds*. The Cu/Zn phase diagram (Fig. 6.16) is alarmingly complicated, but if

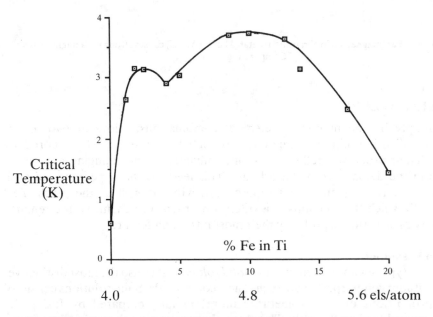

Fig. 6.15 — Variation of superconducting transition temperature with composition in Ti/Fe
alloys. Redrawn from Matthias, B. T. in Tanenbaum, M. & Wright, W. V. (1962) *Superconductors*. Interscience, New York.

Box 12 Superconductivity
   When many metals and alloys and some other materials (Chapter 11) are cooled to near 0 K, their electrical resistance drops abruptly to zero. They become *superconducting*. The temperature at which this takes place is known as the critical temperature (T_c) or the superconducting transition temperature. Among the elements, Nb has the highest T_c, 9.3 K, while the T_c value of 23.4 K for Nb_3Ge is the highest known for metal alloys.
   Superconductivity is discussed in more detail in Chapter 11.

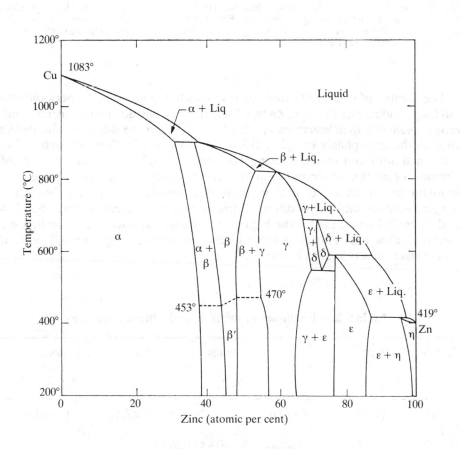

Fig. 6.16 — Phase diagram for 'brass', the alloys of Cu and Zn.

we allow for the inevitable solid solutions (and ignore the high temperature δ phase), we find the phases listed in Table 6.1. The ideal formula Cu_5Zn_8 is rather strange, but the formulae of other alloys that share the same β, γ and ε brass structures are far more bizarre (Table 6.2). Understanding comes when we calculate the ratios of valence electrons to atoms, and find that each phase has a constant electron:atom ratio.

Table 6.1 — Phases of brass

Phase	alpha	beta	gamma	epsilon	eta
	α	β	γ	ε	η
Ideal formula	Pure Cu	CuZn	Cu_5Zn_8	$CuZn_3$	Pure Zn
Structure type	c.c.p	b.c.c.[a]	complex cubic[b]	h.c.p.[a]	distorted h.c.p.

[a] Usually, the different atoms are randomly distributed over the sites of the b.c.c. and h.c.p. structures, but some more complicated arrangements occur. For example, above 470°C, β-brass (CuZn) is disordered, but below that temperature each Cu has eight Zn neighbours and *vice versa* (in the CsCl structure).
[b]The unit cell contains 52 atoms.

The number of valence electrons corresponds to the group in the periodic table, with Cu providing one electron, Zn two, etc., but the transition metals are counted as zero; presumably their electrons are firmly localized in the d-bands. The defining ratios for the three phases are 3:2, 21:13, and 7:4 (Table 6.2). These ratios have been established with considerable confidence, even though the occurrence of solid solutions means that observed compositions often vary over a considerable range. Undoubtedly, structural control in these compounds is exercised through the changing relative stabilities of different structures as their electron energy bands are filled to critical levels (as for the b.c.c./c.c.p. transition discussed in section 6.2). However, it has not yet been possible to discover what energy criteria lead to the specific electron:atom ratios observed.

Table 6.2 — Examples of the β, γ, and ε brass structures

Electron: atom ratios	β-brass 3:2	γ-brass		21:13	ε-brass	7:4
CuZn) CuBe)	(1+3):2	Cu_5Zn_8	(5+16):13	$CuZn_3$		(1+6):4
Cu_3Al)	(3+3):4	Fe_5Zn_{21}	(0+42):26			
				Cu_3Sn	(3+4):4	
Au_3Al)						
Cu_5Sn) Cu_5Si)	(5+4):6	Cu_9Al_4	(9+12):13	Ag_5Al_3		(5+9):8
CoAl	(0+3):2	$Cu_{31}Sn8$) $Na_{31}Pb_8$)	(31+16):39			
$CoZn_3$	(0+6):4					

6.4.3 Alloy structures controlled by metal atom size

The majority of alloy structures exist with more than one combination of metals, but their formulae do not show constant electron:atom ratios. Instead, the allowed combinations depend on the relative sizes of the atoms. This section examines some of the alloys of this type. More examples (with more structural detail) are given in Wells (1984), pp. 1300–10. By combining metal atoms of differing sizes, coordination numbers greater than 12 are possible. Polyhedra with 14, 15, and 16 vertices are particularly common, and are shown in Fig. 6.17a–d. These polyhedra can be linked in a variety of ways.

One characteristic group of size-dominated alloys comprises the Laves phases (Barry & Raynor 1953). These have formula AB_2, with three closely related structures, and are found for at least 220 combinations of metals; the type examples of the three structures are $MgZn_2$, $MgCu_2$, $MgNi_2$. The A atom is rather larger than

 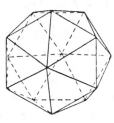 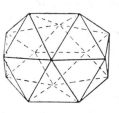

(a) 12   (b) 14   (c) 15

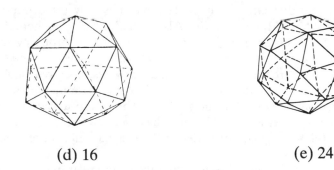

(d) 16   (e) 24

Fig. 6.17 — Typical coordination polyhedra in metal alloys (a) 12: icosahedron (bicapped pentagonal antiprism), (b) 14: bicapped hexagonal antiprism, (c) 15: hexagonal antiprism with 1 and 2 atoms capping the hexagonal faces, (d) 16: truncated tetrahedron of 12 atoms, with 4 atoms capping each hexagonal face (known as the Friauf polyhedron), (e) 24: snub cube, found as the coordination polyhedron of Rb in $RbZn_{13}$. Sources: (a-d), reproduced with permission from Wells (1984), Fig. 29.14, (e) data from Bruzzone, G. (1971) *Acta Cryst* **B27** 862.

B, with an ideal radius ratio of 1.225:1. The A atoms are 16-coordinated by 12 B and 4 A atoms, while the B atoms have 6 A neighbours and 6 B neighbours. The great popularity of this structure probably arises because these interpenetrating polyhedra are neatly arranged in the overall structures, which combine either 4 or 8 A polyhedra with 8 or 16 B polyhedra.

The σ('sigma') phase,† formed by transition metals either as a binary or a ternary phase, is important technologically, because it is brittle. The structure has a tetragonal unit cell containing 30 atoms with 12-, 14-, and 15-coordination. It occurs with a variety of formulae: Ni_9V_{21}, $Ni_{11}V_{19}$, $Ni_{13}V_{17}$, MnMo, $Co_{13}Cr_{17}$, etc. Because these atoms have almost identical X-ray scattering power, structure determination with X-rays cannot decide which atoms have which coordination (or indeed if the structure is disordered). Using neutron diffraction, the atoms can be distinguished, confirming the logical supposition that the larger atoms in the σ-phases (V, Cr, Mn, Mo) tend to occupy the 14 and 15 coordinate sites, the smaller atoms (Mn again and Fe to Ni) the 12 coordinate sites.

With a large discrepancy in metal atom size, coordination numbers as high as 22 and 24 occur, in alloys with a very large A:B ratio, the XY_{11} and XY_{13} structures; examples are $LaZn_{11}$, $LaBe_{13}$ and $RbZn_{13}$. The 24-vertex polyhedron is an elegant 'snub' (truncated) cube, illustrated for $RbZn_{13}$ in Fig. 6.17e. The atoms of the major components in these alloys make contacts with each other at very similar distances to those in the pure metals, so it is easy to visualize them as metals which have distorted just enough to insert a large foreign body into the structure. These structures are currently known only with Be, Cd, and Zn as the major component, but many more examples must surely exist. Rather similar compounds are formed by boron, with such formulae as ThB_4, CaB_6, UB_{12}. They contain linked boron atoms, often in clusters (B_6 or B_{12}) which provide similar metal environments to the XY_{11} and XY_{13} structures. However, these borides are normally regarded as ionic compounds, such as $nCa^{2+}[B_6]_n^{2n-}$, rather than as alloys.

The final example in this section comes from a most surprising group of alloys, those of Mg with Al; these metals have 12-coordinate radii of 160.2 and 144.5 pm, and of course have no d-electrons. The system includes only three intermediate phases, β-Mg_2Al_3, γ-$Mg_{17}Al_{12}$, and ε-$Mg_{23}Al_{30}$ (α and δ being elemental Al and Mg). The simplest formula, Mg_2Al_3, corresponds to what is probably the most complex alloy structure yet discovered, with 1168 atoms in a cubic unit cell (2824pm on each side). The structure follows the pattern already described reasonably well, having 672 atoms in 12-coordinate sites, 252 with 16-coordination, and another 244 less regularly coordinated with 10–16 neighbours. The beautiful pattern in which these coordination polyhedra are linked cannot be described here, but the reader is urged to examine the seven pages of plates in the original publication. (Sampson 1963).

The γ and ε-phases in the Mg-Al system are much more restrained, with only 58 and 53 atoms respectively in their cubic unit cells, but the former is of particular interest. The unit cell contains two formula units, and of the 34 Mg atoms, 24 are 13-

† The naming of alloys is very confusing. The initial identification of a phase is often in a phase diagram, such as Fig. 6.16. Components may then be labelled with a specific (and even logical) Greek letter. Trouble starts when the phase is structurally characterized and recognized in other alloys. The name is often transferred to the new system, modified in some way to avoid (or create) confusion. The σ-phase was first identified in 1927, in the Fe–Cr–Ni system.

coordinate and 10 are 16-coordinate, while all 24 Al atoms are 12-coordinate. The most intriguing aspect is that this is isostructural with the room temperature form of the element Mn (α-Mn). In the latter's unit cell the 24 12-coordinate atoms have Mn-Mn distances of 224–287 pm, while for the 24 atoms with 13-coordination and 10 with 10-coordination the distances are 235–291 pm and 272–291 pm. The comparison with the $Mg_{17}Al_{12}$ phase allows us to understand this structure. Some of the Mn atoms are behaving as if they were larger than the rest — corresponding to Mg^{2+}, rather than Al^{3+}. Bearing in mind the propensity for Mn to show variable valency, we can easily identify these as Mn^{2+}, with the smaller atoms being Mn^{3+}, and understand that Mn takes up this uniquely complicated structure because it is really an alloy of (Mn^{2+} + Mn^{3+} + electrons).

6.4.4 Interstitial compounds

If transition metals are combined with much smaller atoms that are not too electropositive, the results are very different both in structure and in properties from XY_{11} and XY_{13} (which also involve large size differences). It is useful to include the compounds formed between transition metals and C, N, and H under the general heading of alloys, because they retain much of their metallic character. These small atoms occupy the *interstices* in close-packed metal structures, and the compounds are therefore called *interstitial compounds*.

Close-packed structures contains two types of holes, one larger than the other.†
The simplest hole to visualize is formed between one triangle of atoms and the single atom above this triangle in the next layer (Fig. 6.18a). An atom will fit into this hole as long as its radius is no more than $0.225r_m$ (r_m = radius of the metal atom). This hole is surrounded by four metal atoms at the corners of a tetrahedron, and it is therefore described as a *tetrahedral hole*. Each atom in the crystal is associated with eight tetrahedral holes: three between it, two neighbours in the same layer, and an atom of the next layer up; three similarly with the next layer down; one hole between it and a triangle of atoms in the layer above it; one similarly with the layer below. As each hole is bordered by four atoms, in all there are *two* tetrahedral holes per metal atom.

The second type of hole lies between a triangle of atoms in one layer, adjoining another triangle in the next layer (rather than a single atom) (Fig. 6.18b). This hole is rather larger, radius $0.414r_m$. It is surrounded by six atoms with the form of an octahedron. It is therefore an *octahedral hole*. As each atom has six such holes around it, the structure contains one octahedral hole per atom. Taking a typical transition metal radius as about 150 pm, an atom must have a radius less than about 65 pm to fit into an octahedral hole, or less than 35 pm for a tetrahedral hole. Even though the conventional radii of C, N, and H are larger than this, it is found that C and N will fit in the octahedral and H in the tetrahedral holes, without appreciably expanding the metallic lattice.

The preparations of interstitial compounds require fairly extreme conditions: heating metals with carbon to 2200°C or NH_3 (1100°C), or in a N_2 or H_2 atmosphere (1000°C). Exceptionally, Pd will absorb and release H_2 reversibly at room tempera-

---

† As with the metal structures themselves, it is much easier to visualize these holes with models of the structures rather than from pictures.

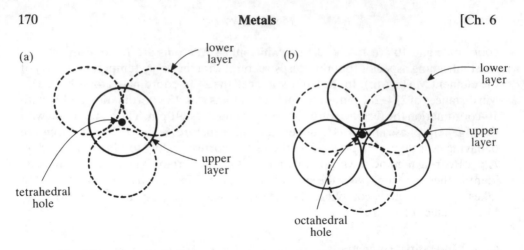

Fig. 6.18 — Interstitial holes in close-packed structures (a) tetrahedral (b) octahedral.

ture. Interstitial compounds are mainly formed with elements from the left-hand half of the transition metals, for reasons that are not clear. Despite their remarkable stability, their heats of formation are low, apparently because of the need to break the strong C–C, N–N, or H–H bonds, and these heats of formation decrease from left to right across the transition series (Phillips & Williams 1965, vol. 1, p.615).

Interstitial compounds invariably show a range of compositions; some ideal formulae are listed in Table 6.3. It is a remarkable property of the hydrides such as

Table 6.3 — Ideal formulae of interstitial compounds

Carbides	Nitrides	Hydrides
TiC	TiN	TiH_2
HfC	ZrN	CrH, CrH_2
W_2C	TaN	ZrH_2
NbC		HfH_2
		PdH

TiH_2 that they contain hydrogen at twice the density of solid H_2. This is the chemical environment to the proposed 'cold fusion' reactions in TiH_2 and PdH, reported in 1989.

Although the description of the structures suggests that the C, N, or H atoms simply occupy the interstices in the metal crystal (the original view taken of these compounds), this is an oversimplification. At their formation temperatures, the metal atoms are mobile, and structural rearrangements take place. In particular, the interstitial compounds almost always have c.c.p. structures, although the parent metals are mostly b.c.c.† Almost the only metal which does not rearrange on forming a hydride is Pd, as it already has a c.c.p. structure. This underlies its unique ability to take up hydrogen at room temperature.

As materials, interstitial compounds retain metallic bonding, metallic magnetic behaviour, and electrical conductivity, while several are superconducting. They are chemically inert and do not react with air or water, even at high temperature. However, their most remarkable properties are their high melting points and their hardness. Melting points include: W_2C, 3030; HfC, 4160; 4TaC/ZrC, 4215; ZrN, 3220°C, and all of these compounds have hardness values of 8–10 on the Moh scale (whose limiting value is 10, the hardness of diamond). These properties, of course, make them extremely useful in applications such as hardened tips for machine tools.

The bonding in interstitial compounds has been the subject of much speculation, suggestions covering every species from C^{4+} to C^{4-} via covalently bound C, and 'no-bond' atoms slipping lithely in and out of the lattice! However, their variable composition and retention of metallic conductivity must place them as extreme examples of alloys, in which the electron-rich C and N atoms are able to introduce electrons into the energy bands of the metals, while at the same time strongly perturbing the band structure because of their electron-attracting ability. The remarkable stability that these perturbations impart is undoubtedly their most interesting feature. It has not received a satisfactory theoretical explanation, though one starting point might be the consideration that the small size of the interstitial atoms leads to a concentration of attracting centres (M and C, N, H) more dense than in any other alloys.

6.5   APPENDIX: THIRD IONIZATION POTENTIALS OF THE LANTHANIDES

The standard electron configurations of the isolated lanthanide atoms are $6s^2 4f^n$ ($n = 0$–14). As Fig. 6.10 shows, their first and second ionization potentials (removal of two s-electrons) vary smoothly with only minor irregularities. For I_3 (the removal of one f-electron), we would expect a similar smooth increase along the lanthanide series, as the nuclear charge increases, but the screening due to the addition of the outer electrons to the same shell (the f-shell) does not rise to the same extent. The graph (Fig. 6.10) does show this increase, but superimposed on it is a sudden step, between Eu and Gd. This arises because of the preference for electrons to be aligned parallel rather than anti-parallel (Hund's second rule) (Box 11). The corresponding energy term is known as the *correlation energy*, and it is the existence of this energy which underlies Hund's rule.

The correlation energy can be written as

$$CE = k.n(n-1)/2$$

---

† W_2C and some other M_2X compounds have h.c.p. metal structures. It is possible to fill half the octahedral holes in a h.c.p. structure, but the remaining empty holes are very close to the filled ones. Thus, if the carbon:metal ratio rises above 1:2, the structures must rearrange to c.c.p.

for n parallel electrons.† As far as I_3 is concerned, the essential consideration is what happens to CE when the third electron is removed. In Table 6.4, the CE values for

Table 6.4 — Correlation energy changes for removal of an f-electron

Element	La	Ce	Pr	Nd	Pm	Sm	Eu	Gd	Tb	Dy	Ho	Er	Tm	Yb
$f^{n+1}(2+)$	1	2	3	4	5	6	7	8	9	10	11	12	12	14
CE (2+)	0	1	3	6	10	15	21	21	22	24	27	31	36	42
$f^n(3+)$	0	1	2	3	4	5	6	7	8	9	10	11	12	13
CE (3+)	0	0	1	3	6	10	15	21	21	22	24	27	31	36
δ(CE)	0	1	2	3	4	5	6	0	1	2	3	4	5	6

the f^n states of the M^{3+} and the f^{n+1} states of the M^{2+} ions are compared. To remove the third electron, the appropriate electrostatic energy has to be supplied (increasing smoothly along the series). In addition, any decrease in CE has to be made up. Moving along the series up to Eu, this loss in CE builds up as the number of parallel electrons increases. However, Gd^{3+} has eight f-electrons. Seven point in one direction and provide 21 units of CE, but the eighth is parallel to no other electrons. Its removal therefore needs no extra CE. I_3 for Gd is lower than that predicted by extrapolation from Eu by about 500 kJmol^{-1}.

As well as the main step where the f-shell is half full, the plot of I_3 shows two minor irregularities, at the 1/4 and 3/4 positions in the shell. These are also related to Hund's rules, not to the second but the third rule, which requires that electrons have maximum orbital angular momentum. The total orbital angular momentum (L) for a given electron arrangement is easily calculated by adding the values for each occupied orbital (Fig. 6.19). These values are normally stated as letter equivalents, S,P,D,F,G,H,I for L=0,1,2,3, etc. Thus, Fig. 6.19b-e show I- and H-states, and Fig. 6.19f summarizes the values. For ionization of f^1, f^2, f^3, f^8, f^9, and f^{10} states, orbital angular momentum is lost on ionization, and so I_3 is increased by the extra energy needed. For f^4 and f^{11}, L does not change (Figs 6.19b–c), but for f^5, f^6, f^7, f^{12}, f^{13}, and f^{14}, it actually increases on ionization. Thus, the elements Pm, Sm, Er, and Tm have I_3 values lower than expected. This variation is largest for the elements with highest L-values, i.e. Pr, Sm, Dy, and Er.

These effects are summarized in Fig. 6.20, which can be compared with the plot of I_3 for the lanthanides in Fig. 6.10. Precisely the same effects operate for the transition metals (Fig. 6.12), though the 1/4 and 3/4 shell effects are small (because of the lower values of L). The variations in I_3 have a profound effect on the chemical behaviour of the lanthanides, which is discussed by Johnson (1977).

---

† This is more usually presented in terms of what are known as Racah parameters, relating to other aspects of electron interaction as well as correlation energy. See Johnson,D.A. (1969) *J. Chem. Soc. (A)*, p. 1525.

(a) orbital angular momentum
   of each orbital

+3	+2	+1	0	−1	−2	−3

(b) f^3:Nd^{3+}

L = 6, I − state

↑	↑	↑				

(c) f^4:Pm^{3+} or Nd^{2+}

L = 6, I − state

↑	↑	↑	↑			

(d) f^{11}:Er^{3+}

L = 6, I − state

↑↓	↑↓	↑↓	↑↓	↑	↑	↑

(e) f^{12}:Tm^{3+} or Er^{2+}

L = 5, H − state

↑↓	↑↓	↑↓	↑↓	↑↓	↑	↑

(f) Summary

f^n	0	1	2	3	4	5	6	7
state	S	F	H	I	I	H	F	S
f^n		8	9	10	11	12	13	14
state		F	H	I	I	H	F	S

Fig. 6.19 — Orbital angular momenta for lanthanides (a) values for each orbital (b)-(e) values for f^3, f^4, f^{11} and f^{12} (f) summary.

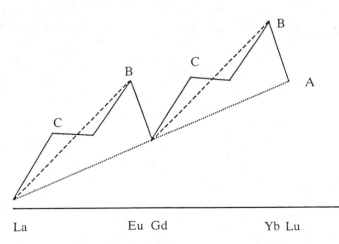

Fig. 6.20 — Summary of variations in I_3 for lanthanides. A: basic trend of increasing nuclear charge. B: loss of correlation energy, M^{2+} to M^{3+}. C: loss or gain of orbital angular momentum, M^{2+} to M^{3+}.

FURTHER READING

Theory and properties of metals
Bube, R. H. (1981) *Electrons in solids*. Academic Press, New York, Chapter 7.
Coles, B. R. & Caplin, A. D. (1976) *Electronic structure of solids*. Edward Arnold, London, Chapter 4.
Cox, P. A. (1987) *The electronic structure and chemistry of solids,* Oxford University Press, Oxford.

Total metallic bond enthalpy (T.M.B.E.)
Alcock, N. W. (1990) *Chem. Phys. Lett.* **165** 339. The correlation of TMBE and metallic radius was discovered during the writing of the present book, and it is not yet clear why these plots should be linear.

Alloy structures
Wells, A. F. (1984) *Structural inorganic chemistry*. Clarendon Press, Oxford (5th ed.)

Interstitial compounds
Phillips, C. S. G. & Williams, R. J. P. (1965) *Inorganic chemistry,* Clarendon Press, Oxford, vol. I.

Lanthanides
Johnson, D. A. (1977) *Adv. Inorg. Chem. Radiochem.* **20** 1.

Text references
Berry, R. L. & Raynor, G. V. (1953) *Acta Cryst* **6** 178. (Gives full details of the Lavers phase).
Sampson, S. (1963) *Acta Cryst.* **19**, 401.

7

The accompaniment of covalent bonding

7.1 INTRODUCTION

This chapter looks at a variety of bond types with one feature in common: that they co-exist with covalent bonds. The most universal of these is *van der Waals bonding*, the weak attraction that holds atoms of all sorts together. However, the most significant, judged by the range of compounds in which it is important, is certainly *dative* or *donor-acceptor bonding*. The general examination of dative bonding is followed by a review of two specific varieties: *hydrogen bonding* and *secondary bonding*. Their distinctive character is related to the particular classes of compound in which each is found. The most widely distributed type of dative bonding of all — that found in transition metal complexes — is the subject of Chapter 8.

7.2 VAN DER WAALS BONDING

Van der Waals bonding is the general name given to the weak forces that exist between all atoms. These forces make possible, for example, the condensation of the noble gases to liquids. They are particularly significant in covalent solids as an essential part of their bonding. Apart from diamond and the few other compounds with infinite 3-dimensional structures, van der Waals forces hold the molecules of all covalent solids together. They are also important *within* molecules, leading to preferences for one geometry over another.

   Van der Waals forces are electrostatic in origin, even when the molecules concerned do not carry a charge. They arise in three ways. The simplest to visualize are the forces between molecules that have non-uniform charge distributions, dipoles (Fig. 7.1a) or more complicated arrangements in molecules with more than two atoms (Fig. 7.1b). These forces are known as *Keesom* forces. Such molecules will obviously attract each other as long as they are oriented suitably. Fig. 7.1c shows how the attraction varies with orientation. Although the molecules try to take up favourable positions, their thermal energy tends to disorient them. As a result, the

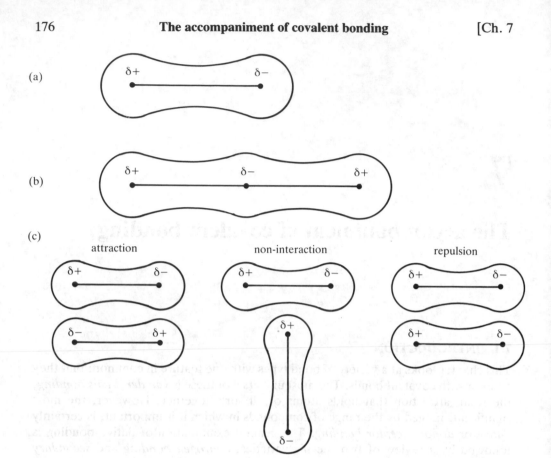

Fig. 7.1 — Molecules with non-uniform charge distribution. (a) a dipolar molecule, (b) a quadrupolar molecule, (c) attractive, neutral, and repulsive orientations of two dipolar molecules.

formula for the attractive energy of two dipolar molecules (7.1) depends inversely on the temperature

$$E_k = -2\mu_1^2 \cdot \mu_2^2/3(4\pi\varepsilon_0)^2 kTr^6. \tag{7.1}$$

The energy depends on the two dipole moments (μ_1 and μ_2), and most importantly on $1/r^6$, as well as the temperature. This $1/r^6$ dependence shows that these forces fall off rapidly with distance.

The second component of the van der Waals forces arises when a molecule with a dipole approaches an atom or molecule *without* one. Fig. 7.2 shows that the positively charged end of the dipole attracts the electrons of the other atom. This gives it an *induced dipole* and causes an attraction between the two atoms, the *Debye* force. Its energy (E_d) is given by (7.2)

$$E_d = -2\alpha\mu^2/(4\pi\varepsilon_0)^2 r^6 \tag{7.2}$$

Fig. 7.2 — A dipole on one molecule induces a dipole on a non-dipolar atom.

Again this energy depends on $1/r^6$ and the dipole moment μ, and also the polarizability of the second atom α, which controls the size of the induced dipole. However, the temperature is not involved because the attractive force is independent of the orientation of the dipole.

It is less easy to see why any force exists between two atoms, neither of which has a dipole, e.g. the atoms of a noble gas. In fact, the third type of van der Waals force, known as the *London* force, is found between all atoms, whether or not they possess dipoles. This force depends on the quantum-mechanical properties of electrons, rather than on simple electrostatic attraction. Electrons do not merely make up a smeared-out sphere of negative charge, but move around the atom. As a result, at any given time, their distribution is irregular and the atom has a temporary dipole (Fig. 7.3). This, of course, induces a dipole on an adjacent atom and causes an

temporary dipole   induced
on atom 1   dipole on atom 2

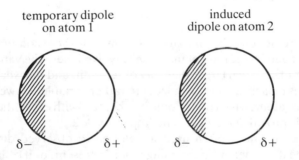

Fig. 7.3 — A temporary dipole on one atom induces a dipole on atom 2.

attraction. The corresponding energy $(E_1$; equation 7.3) depends on the polarizability of the atoms and the frequency ν_0 at which the electrons oscillate, which can be calculated by quantum mechanics

$$E_1 = -3h\nu_0\alpha^2/4(4\pi\varepsilon_0)^2 r^2. \tag{7.3}$$

Like all other bonding forces, the attractive van der Waals forces are also accompanied by repulsive forces. Their precise calculation is difficult, but they are often treated as varying as k/r^{12}, with the constant k obtained from experiment. Some numerical values for the van der Waals energies are given in Table 7.1. It is worth noting how small they are, unless the molecules have substantial dipoles.

Table 7.1 — Van der Waals energies (kJ mol^{-1})

Compound	London energy	Keeson energy	Debye energy	Total energy	Observed ΔH_{subl}	B. Pt/K
Ar	−8.5	−0	0	−8.5	−8.4	76
H$_2$	−0.8	−0	0	−0.8	−0.8	20
CO^a	−8.4	−0	(0.002)	−8.4	−7.9	81
HCl	−16.8	−3.3	−1.0	21.1	−20.0	188

^{a}CO has an extremely small dipole moment, so the Debye energy is very small and the Keeson energy smaller still.
Source: Day & Selbin (1969), p. 177.

The structural effects of van der Waals bonding are just what we would expect for an attractive force that does not depend on direction. The noble gases crystallize with cubic close-packed structures, with h.c.p. an alternative for Ar. Other molecules, such as H$_2$ and N$_2$, have similar structures in the solid, with rotation making the molecules appear spherical.

7.2.1   Van der Waals radii
As with other interactions, the interatomic distances in these non-bonded attractions can be characterized by the radii of the atoms involved, here the van der Waals radii. These are, of course, much larger than any other radii, and the weakness of the van der Waals forces means that the distances are rather variable. However, radii can be assigned to those atoms which often come into contact with each other. Careful study by Bondi (1964) provided the figures in Appendix A.2.

These values are most often used in relation to observed distances between atoms, to decide if van der Waals bonding alone is present, or if some stronger bond is likely. It is often convenient to apply a simple approximation in place of a precise calculation: that a van der Waals bond is about 150 pm longer than the corresponding single bond. For example, Fig. 7.4 shows the structure of the complex between ICl

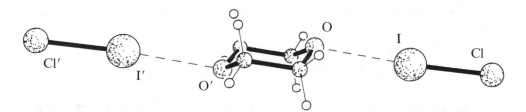

Fig. 7.4 — Structure of the ICl–dioxan complex, with I . . . O distance of 257 pm. Source: data in Hassel, O. & Hvoslef, J. (1956) *Acta Chem. Scand.*, **10** 138.

and dioxan, in which the molecules are arranged in a chain, with I–O distances of 257 pm. Comparison with the sum of the van der Waals radii for I and O (350 pm) shows that the I–O interaction must be much stronger than a van der Waals force.

7.2.2   Van der Waals forces in molecular mechanics

Van der Waals forces find an important place in molecular mechanics (see section 2.3.2). The energies of non-bonded interactions cannot be neglected in calculating the overall conformational energy. As well as the attractive energy, the repulsive component is important in preventing non-bonded atoms from coming unreasonably close together — or rather giving the energy cost of a distortion that may be favoured by other forces.

These methods have also been successfully applied to predicting how molecules will be arranged in crystals, when only van der Waals forces are involved. It would seem that with energy $\sim 1/r^6$, only the contacts between molecules and their nearest neighbours need be considered. However, this neglects the fact that the number of adjacent molecules increases as r^3, so the overall fall-off in energy with distance is slow. As an example (Burkert & Allinger 1982), to achieve satisfactory agreement with experiment for the sublimation energy of n-C_6H_{14}, it was necessary to consider a block $15 \times 15 \times 15$ molecules — 3375 in all — a very lengthy calculation, even though it dealt with only 10^{-20} mole!

7.3   DONOR–ACCEPTOR BONDING

Donor-acceptor or dative bonding is a form of covalent bonding between two molecules that themselves exist independently. A characteristic example is

$$NMe_3 + BCl_3 \quad \rightarrow \quad Me_3N \rightarrow BCl_3$$

The product $Me_3N \rightarrow BCl_3$ is known as an *adduct*, and the arrow in its formula indicates the dative bond. The two components of the adduct are a *donor*, NMe_3, which has an unshared pair of electrons, and an *acceptor*, BCl_3, which has a vacant orbital. A molecular orbital diagram of the bonding is shown in Fig. 7.5. The bonding

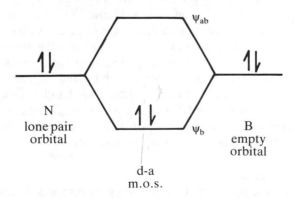

Fig. 7.5 — Molecular orbital diagram for the formation of the dative bond in $Me_3N \rightarrow BCl_3$.

molecular orbital ψ_b is formed from both parent orbitals, and therefore becomes more associated with the boron than the original lone pair orbital on the nitrogen. Electron density therefore moves from nitrogen, the donor, towards boron, the acceptor.

It is also possible to describe donor-acceptor bonding as simple covalent bonding between charged atoms, i.e.,

$$NMe_3 + BCl_3 \quad \rightarrow \quad Me_3N^+ \cdot + BCl_3^- \cdot \quad \rightarrow \quad Me_3N^+ - {}^-BCl_3$$

In the first stage, one electron is transferred from the donor to the acceptor. Each molecule then corresponds to a carbon atom forming three bonds, with one unpaired electron available for a fourth bond. This description is particularly useful when either the donor or the acceptor is itself charged, e.g.

$$NH_3 + H^+ \quad \rightarrow \quad NH_4^+$$
$$BCl_3 + Cl^- \quad \rightarrow \quad BCl_4^-$$

Because there is no distinction between the four H's or Cl's, these ions are easily regarded as equivalent to the corresponding carbon compound with a tetravalent but charged central ion.

Fig. 7.5 conceals one important aspect of dative bonding for BX_3 compounds, such as BCl_3. These are planar molecules, sp^2 hybridized with the 'empty' fourth orbital on B of pure p-type. This orbital overlaps with the full p orbitals on each chlorine, to give a π-molecular orbital system (Fig. 7.6a, as described in section 5.9.4). For donation to take place, the BCl_3 has to adopt tetrahedral geometry (Fig. 7.6b), converting the empty π_{ab} orbital into an sp^3 hybrid and destroying the π molecular orbital system.

An intriguing contrast arises from this. BF_3 with its 24 valence electrons is isoelectronic with NO_3^-, CO_3^{2-}, and $BO3_3^{3-}$. The two boron compounds show acceptor properties and will give tetrahedral $[BF_3X]$ and $[BO_4]$ groups (with the O-atoms linked to other groups). In contrast, NO_3^- and CO_3^{2-} have no acceptor ability at all, even though they are identical in electronic structure to BF_3. The difference lies in the relative electronegativities of the central and the terminal atoms. The strongly electron-attracting fluorine or oxygen atoms reduce the degree of interaction between the outer orbitals and the central orbital, leaving the boron still able to accept additional electron density. In NO_3^- on the other hand, the electron attraction by N leaves the electron density at N much higher. The π_{ab} orbital is now very high in energy. Rearranging the molecule to allow this orbital to become an sp^3 hybrid able to accept an electron pair has a prohibitive energy cost.

7.3.1 Alternative names

Two other names are often given to donor-acceptor complexes. The first is *charge-transfer complex*, applied especially to weakly-bound complexes, such as ICl-dioxan (Fig. 7.4). This relates to the formation of the donor-acceptor molecular orbitals.

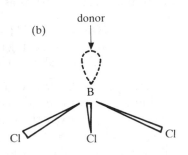

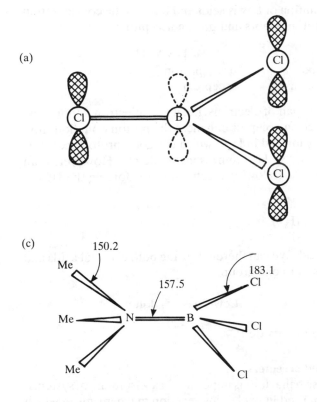

Fig. 7.6 — BCl₃ (a) p-orbitals in the molecule, (b) empty boron orbital in a tetrahedral donor-acceptor complex, (c) geometry of the complex with Me₃N.

Promotion of an electron from the ψ_b to the ψ_{ab} orbital (Fig. 7.5) produces an absorption band that is absent in the parent compounds. Because the electron originates on the donor and moves to the empty orbital originally on the acceptor, this corresponds to electron transfer from donor to acceptor, and the absorption is therefore called a *charge transfer band*. By extension, the complexes are *charge transfer complexes*. The charge transfer band is generally in the near ultraviolet, sometimes running into the blue end of the spectrum which gives the complexes a yellow colour. The change in colour of I_2 solutions from blue in $CHCl_3$ to brown in C_2H_5OH is due to a charge-transfer absorption by an I_2–C_2H_5OH complex.

Donors and acceptors of all types are also described as *Lewis bases* (donors) and *Lewis acids* (acceptors), after G. N. Lewis, who first proposed these names in 1923. These are much more complicated terms and confusing in some ways; they must be clearly distinguished from ordinary aqueous acids and bases. Even though the discussion of donors and acceptors might be simpler without these names, the concepts underlying them are so important and the names so widely used, that they cannot be ignored.

The starting point for the definition of Lewis acids and bases is the consideration of ordinary acids and bases as proton donors and proton acceptors:

$$HX \rightarrow X^- + [H^+] \qquad\qquad Y: + [H^+] \rightarrow YH^+$$

acid (needing to be   base (proton

    taken up by a base)   acceptor)

Obviously a base has to have a lone pair of electrons, i.e. it is an electron donor. Less obviously, the presence of an electron acceptor enhances the proton donation from an acid. HCl, for example, is fully ionized in H_2O which is a good proton acceptor. Nitrobenzene is a poor acceptor and so HCl ionizes incompletely. However, if an electron acceptor, $AlCl_3$ is added, the following reaction occurs, forcing the HCl to ionize further.

$$AlCl_3 + HCl \rightarrow [AlCl_4]^- + [H^+]$$

The $AlCl_3$ functions as a sort of acid. We can therefore bring both classical acids and bases, with donors and acceptors, into one group:

    Lewis acid $\equiv$ acceptor   Lewis base $\equiv$ donor

(and the qualification 'Lewis' is often omitted).

7.3.2 Classification of donors and acceptors

Donor/acceptor or Lewis acid/base behaviour is important in a vast range of systems. For example, the concept is often used in explaining reaction mechanisms in which two molecules are associated in the transition state. Carbonium ions and carbanions are respectively Lewis acids and bases, and therefore react with other molecules of complementary character. Our present concern is however with stable donor-acceptor systems rather than with intermediates.

  Given the two requirements: a lone pair and an empty orbital, we can assemble the main types of donors and acceptors (Table 7.2). The complexity of this table indicates how pervasive donor-acceptor bonding is. We can distinguish three major categories, depending on the strength of the bonding.

 (i) An enormous number of weakly bound complexes can be identified by the spectroscopic changes on their formation. They are difficult to isolate as solids, and when isolated have long distances between donor and acceptor atoms. These are often described as charge-transfer complexes. A classic example is the ICl-dioxan complex (Fig. 7.4). The intriguing ability of aromatic complexes to function both as donors and acceptors demonstrates the flexibility of this type of interaction.

 (ii) Acceptors in groups 1–3 in Table 7.2 form fairly strongly bound complexes which can usually be isolated. Their donor-acceptor bonds are similar in length to covalent bonds, e.g. $Me_3 \rightarrow BCl_3$ (Fig. 7.6).

(iii) Metal cations form complexes with donors ('ligands'). These are considered in the next chapter.

Table 7.2 — Types of donors and acceptors (with some examples of complexes given in [ ])

Donors: *with lone pair*

A — σ-donors

1. Group 15 atoms with valency 3

   NH_3, NR_3, C_5H_5N, PR_3

2. Group 16 atom with valency 2

   H_2O, $(C_2H_5)_2O$, $(CH_3)_2S$, C_5H_5NO (pyridine N-oxide, donating from O)

3. Group 17 atoms with valency 1

   X^- (halide ions) (R– X are poor donors)

4. Other anions and a few related species

   CO, NO, O_2, H^-, CN^- (donating from C or N)

B — Pure π donors

5. Molecules with bonding π-electrons

   C_2H_4, C_6H_6, other aromatics

C — Combined σ donors or acceptors and π acceptors or donors (section 7.3)

6. Donor molecules (class 4 above) with vacant π_{ab} orbitals

7. Transition metal atoms with filled d-orbitals for donation and vacant orbitals for acceptor properties [$Ni(CO)_4$]

Acceptors: *with vacant orbitals*

1. Group 13 atom in covalent molecule

   BF_3, BMe_3, B_2H_6 (as monomer BH_3) [$H_3B{\leftarrow}NMe$]

2. Main group atom (usually Group 14) in low oxidation state

   $SnCl_2$; also O-atom [R_3NO]

3. Atom (2nd row or later) with vacant d-orbitals

   (i) σ-acceptors (orbitals usually hybridised)
       (a) Group 14: $SiCl_4$ [$SiCl_4(NR_3)_2$], Me_3SnCl
       (b) Groups 15–18 atoms in higher oxidation states: PF_5, XeF_6
   (ii) d_π acceptors. See Section 10.4

4. Molecules with empty antibonding orbitals
   (i) σ_{ab} when electron attracting atoms are present

   I_2 [I_3^-, $C_6H_6{\rightarrow}I_2$], $C_2(CN)_4$ (tetracyanoethane, TCNE) [$C_6H_6{\rightarrow}$TCNE]
   (ii) aromatics with low energy π_{ab} orbitals

   $C_6H_3(NO_2)_3$, $C_6H_2)_3OH$ (picric acid)[naphthalene-picrate]
   (iii) donor molecules and atoms of classes 6 and 7 above

5. Cations

   (i) H^+ [NH_4^+]
   (ii) metal cations, especially of transition metals [FeF_6^{3-}]

The distinction in type of orbital involved (between groups A, B, and C in Table 7.2) is unimportant for classes (i) and (ii), where a single orbital of either σ or π type acts as donor or acceptor. Group C involves π orbitals as an addition to a donor-acceptor complex formed with σ orbitals. This extra donation may be in the same or the opposite direction to the σ-donation, and has important effects on the bonding (see Chapter 8).

7.3.3 Donor-acceptor bond strength

Determining the strength of a dative bond is often difficult, but understanding how this strength is controlled by the chemical properties of the combining molecules causes even more problems. This has been the subject of heated arguments that remain unresolved, though the field of battle is quiet at present, probably because the contestants have exhausted the available evidence. The experimental problem is a consequence of the rather low bond strengths (enthalpies) of many donor-acceptor bonds. This has two results. Because of the influence of entropy, equilibrium constants may show a different ordering from the bond enthalpies. Thus, $Me_3N \rightarrow BMe_3$ has a stronger bond than $Me_3P \rightarrow BMe_3$ (73.6 and 69.0 kJ mol^{-1}) but dissociates more readily in the gas phase. Secondly, measurements are often made in solution. Unless the solvent is chosen with great care, it may well form d–a bonds with one of the two components, as well as solvating the product. If the energy of these interactions is not properly taken into account, the apparent bond enthalpy can be very different from the true value; the chemical literature contains many erroneous figures for this reason. Of course from another viewpoint, these solution results are the most useful in practice, even though they differ from ideal predictions. It is also easy to make faulty comparisons. Thus, the boron halides interact with pyridine in the order

$$BI_3 \simeq BBr_3 > BCl_3 > BF_3$$

(because of the energy penalty for losing the π–π interaction; p. 187). However, the BF_4^- ion is much more stable to dissociation than the BCl_4^- ion. We can easily overlook that BF_4^- involves donation from F_4^- and BCl_4^- from Cl_4^-, so this comparison does not involve the same type of interaction as with pyridine.

The most important and remarkable feature of these d–a interaction strengths is that they are *not additive*, in contrast to the various distance and energy terms we have previously examined. This appears from many examples of pairs of donors and acceptors. We can take as donors Et_2O and Me_2S, with acceptors Me_3Ga and I_2. To Me_3Ga, the d–a bond energies are 39.7 kJ mol^{-1} and 33.4 kJ mol^{-1} with Et_2O and Me_2S respectively. We would expect that $Et_2O \rightarrow I_2$ would be stronger than $Me_2S \rightarrow I_2$, but experimentally the enthalpies are 17.6 and 32.6 kJ mol^{-1}. Of course, not every set of energies is reversed in this way. Me_3N forms stronger d–a bonds than Et_2O or Me_2S to both Me_3Ga and I_2. The implication is that it is *impossible* to set up an order of donor and acceptor strength in which a 'stronger' donor will always displace a 'weaker' donor; Et_2O will displace Et_2S from Me_3Ga but the reverse occurs with I_2.

7.3.3.1 *Pearson approach to donor-acceptor bond strength*

The argument over interpretation of d–a bonds lies essentially between a wide-ranging qualitative classification which is broadly (if uncritically) accepted, and a quantitative calculation which reproduces the energies very precisely, but is limited in its scope. Qualitatively, the character of donors and acceptors has been described by R. G. Pearson (1963) in terms of their 'hardness' or 'softness', using the labels 'hard' (or 'soft') acid (acceptor) and base (donor). Acids are separated according to their behaviour with bases. If the stability of their complexes with different donor atoms is in the order

$$N \gg P > As > Sb > Bi; \quad O \gg S > Se > Te; \quad F \gg Cl > Br > I \tag{1}$$

then they are 'hard'. If the stabilities are in the order

$$N \ll P < As < Sb < Bi; \quad O \ll S \sim Se \sim Te; \quad F < Cl < Br \ll I \tag{2}$$

they are soft.

The key principle of this classification is that *hard acids are more stable in complexes with hard bases*, and *soft acids are more stable with soft bases*. In contrast to the graduated scale of base hardness (decreasing in the order (1) above), the acids are grouped as either hard or soft, apart from a few marginal cases (Table 7.3). The

Table 7.3 — Examples of hard and soft acids

Hard	Borderline	Soft
H^+, Li^+–K^+	Fe^{2+}–Cu^{2+}	Cu^+, Ag^+, Au^+, Cs^+
Be^{2+}–Sr^{2+}, La^{3+}	Pb^{2+}, BMe_3	Cd^{2+}, I_2, BH_3
Fe^{2+}, Al^{3+}		$C_6H_3(NO_2)_3$
BF_3, $AlMe_3$, $GaMe_3$		low-valent metal atoms

classification is essentially the result of observation, though systematic trends can be seen. Increased oxidation state causes ions to become harder, as does the attachment of electron withdrawing groups (e.g. BF_3 vs BH_3). The terms 'soft' and 'hard' themselves are intended to indicate whether the ions are easy or difficult to polarize. This property is certainly correlated with acceptor behaviour, but other factors are also important, particularly the ability to accept or donate π-electrons (section 8.3.2).

Various attempts have been made to produce scales of 'hardness' and 'softness', to predict the energies of d-a bonds, but without great success (Pearson 1973). The identification of hard and soft character does allow for bond energy inversions, predicting:

but

$$P_{hard} \rightarrow X_{hard} > P_{hard} \rightarrow Y_{soft}$$

$$Q_{soft} \rightarrow X_{hard} < Q_{soft} \rightarrow Y_{soft}$$

Thus it is consistent with the experimental evidence for these inversions.

7.3.3.2   Drago calculations of donor-acceptor energy
An entirely different approach to the energies of d–a complexes has been made by R. S. Drago (1973). He has shown that energies can be calculated if each acid and base is assigned *two* parameters, E_a, C_a and E_b, C_b. The dative bond energy in the gas phase is then given by

$$\Delta H_{ab} = E_a E_b + C_a C_b \qquad (7.4)$$

This equation predicts energies remarkably precisely (mean error $0.8 \, kJ \, mol^{-1}$ in values of about $30 \, kJ \, mol^{-1}$), for adducts formed between about 50 bases and 30 acids (including about 10 acids forming hydrogen-bonded complexes) (Drago 1973). The values of E and C are not absolute, as four E, C values for two acid/base pairs must be fixed at arbitrary values.

This excellent agreement makes it worthwhile to look carefully at the few misfits. Thus, BMe_3 gives good agreement for its adducts with NH_3 and $MeNH_2$, but with NMe_3, the observed value is about $28 \, kJ \, mol^{-1}$ less than expected. This is plausibly attributed to steric effects. The corresponding compound Me_3C–CMe_3 should be affected by repulsion between its methyl groups to much the same extent. Indeed its heat of formation is reduced by almost the same amount from that predicted by addition of the individual bond enthalpies.

Equation (7.4) is clearly very important in understanding dative bond energies, and any scales that are not in accord with it cannot successfully predict these energies. However, it is limited in two ways. The most disappointing aspect is that it has been applied to only a limited range of acids and bases, mainly forming rather weak acids. It would be interesting to discover if it can be applied to transition metal complexes and other strongly bound donor-acceptor systems. The second problem concerns the E and C values themselves. The E-terms have ben labelled 'electrostatic', and the C-terms 'covalent', but this does not correlate well with their numerical values. However, because these depend on the arbitrarily chosen initial parameters, they are ultimately empirical. Without a theoretical basis for these initial values, it is not possible to begin to interpret the chemical significance of the E and C values.

7.3.3.3 *Bond length and bond strength*

The alternative to direct measurement of bond enthalpy is to examine the bond lengths. This gives a clear overall picture of the pattern of dative bond strengths, though it lacks the precision of the direct method.

The most distinctive type of donor-acceptor complex contain weak interactions, typified by that in ICl . . . dioxan (Fig. 7.4), whose energy is $31.4\,kJ\,mol^{-1}$ with I . . . O distance 257 pm. Comparison with the sum of the single I and O covalent radii (205 pm) shows that this bond is weak, but it is much shorter than would be expected just for a van der Waals interaction (about 350 pm).

In a very different category is the B–N distance of 157.5 pm in $Me_3N{\rightarrow}BCl_3$, which can be compared to the B–C distance of 158 pm in $B(CH_3)_3$. Clearly, the B–N dative bond is virtually equivalent to a covalent bond of order 1.0. A similarly strong bond is found in the amine oxides $R_3N{\rightarrow}O$ with N–O of 140 pm, the same distance as in $[H_3N–OH]^+$. These compounds are sometimes written $R_3N{=}O$, even though the nitrogen does not have five valence orbitals available. Comparison with a typical $R_2C{=}O$ distance (122 pm) shows a great difference between the C–O double bond and the N–O bond (single). In contrast, $R_3P{=}O$ compounds can be described with P–O double bonds; they therefore have significant differences in their chemical properties from amine oxides (see section 9.3.2).

The lengths of dative bonds between boron and halogen atoms are also informative, especially when combined with bond energies. In BF_3, the B–F distance is 131 pm, showing that the B–F bond in BF_4^- (141 pm) is weaker. This is related to the change in shape from planar BF_3 to tetrahedral BF_4^-, and the loss of π-bonding on forming the fourth bond. This helps to understand the ordering of the dative bond strengths of the boron halides with, for example, pyridine ($BBr_3{>}BCl_3{>}BF_3$). This is not what would be expected at first sight. Electron attraction by the halogen atoms should enhance the ability of boron to form dative bonds, making BF_3 the most effective acceptor. Because the experimental acceptor strengths are in the reverse order from this prediction, it is clear that the dominant influence is the loss of π-bonding when the adduct is formed. This is greatest for BF_3, which has the best π–π overlap between boron and halogen p-orbitals.

7.4 HYDROGEN BONDING

Weak intermolecular bonding and dative bonding come together in the next two sections: *Hydrogen bonding* and *Secondary bonding*. Hydrogen bonding in particular is vitally important as a component of the bonding between polar organic molecules, including most of those found in living organisms. It was first proposed as an explanation for the unexpected behaviour of NH_3, H_2O, and HF, in comparison to PH_3, H_2S, and HCl. The boiling points of these first-row hydrides are up to 150° higher than the corresponding second-row compounds. We can see that the effect is related to the presence of the hydrogen atom by comparing the boiling points of H_2O (100°C), F_2O (-224°C), and $(CH_3)_2O$ (-138°C).

Study of the crystal structures of compounds containing hydrogen bonds shows that the bond has a characteristic geometry. Two non-hydrogen atoms, for example two O-atoms in crystals of ice, are rather closer together (276 pm) than the van der Waals' radii would suggest (304 pm). If the hydrogen atoms are precisely located,

e.g. by neutron diffraction (section 2.2.2), they are found to lie close to the line
between the two H-bonded atoms. Fig. 7.7 shows typical dimensions, though the

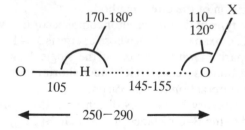

Fig. 7.7 — Dimensions of typical O–H ... O hydrogen bond.

distances and angles are very variable and the O–H . . . O angle may be as low as 160°.
With only X-ray evidence, when the H-atom position may be inaccurate or even
unknown, the short overall O . . . O distance can be used to identify a likely hydrogen
bond.

Spectroscopy can also be used to identify hydrogen bonds. The formation of an
O–H...O bond causes a dramatic shift in the infrared spectrum. The O–H
stretching frequency moves from about $3500\,\mathrm{cm}^{-1}$ to $2000\,\mathrm{cm}^{-1}$, at the same time
becoming much broader and more intense. As would be expected, ^{1}H n.m.r. spectra
are also strongly affected. The chemical shift of the hydrogen moves down-field by up
to 6 p.p.m. The observation of these shifts when a potential H-bond acceptor is
added to a potential donor is particularly useful for identifying weak H-bonds, for
example from H_2S or PH_3.

Asymmetrical O–H . . . O bonds, like that in Fig. 7.7, are the normal type found
in intermolecular contacts. However, geometrical constraints force some hydrogen
bonds *within* molecules to be very short. The shortest example known is in nickel
ethyl-methyl-glyoxime (Fig. 7.8a), which has an O . . . O distance of 240 pm. Short

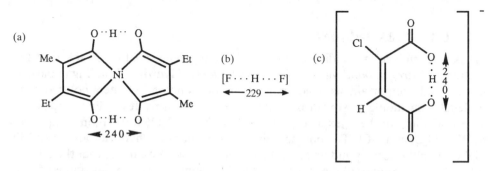

Fig. 7.8 — Short hydrogen bonds (a) Ni-ethylmethylglycoxime (Bovers, R. H., Banks, C. V. &
Jacobson, R. A. (1972) *Acta Cryst.* **B28** 2318), (b) the bifluoride or acid fluoride ion, K[HF₂]
(Ibers, J. A. (1964) *J. Chem. Phys.*, **40** 402), (c) acid salt of chloromaleic acid, K[H(chloroma-
leate)] (Ellison, R. D. & Levy, H. A. (1965) *Acta Cryst.* **19** 260).

bonds are also found in 'acid' salts (Figs. 7.8b,c). Neutron diffraction has shown that in these bonds the H-atom is symmetrically placed. As they lengthen, the H-atom becomes associated with one of the two bonded atoms X–H . . . Y, or X . . . H–Y. At an intermediate stage it is difficult to decide from the diffraction data if the proton is oscillating between two alternative sites, or is moving in a shallow potential well, spending half its time at each end. Fig. 7.9 shows the potential wells for different types of hydrogen bond, illustrating these possibilities.

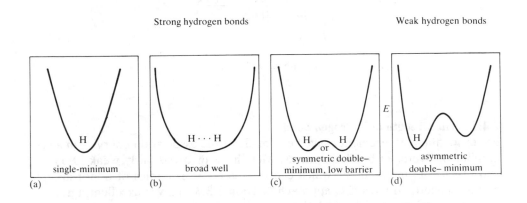

Fig. 7.9 — Potential wells for protons in hydrogen bonds (a) short symmetrical bond, (b) a broad well, or (c) oscillation between two sites in a slightly longer bond, (d) a long bond, showing a possible higher energy (tautomeric) position for the hydrogen atom. Based on: Emsley, J. (1980) *Chem. Soc. Rev.*, **9**, 91.

Classic hydrogen bonds involve N, O, and F as X and Y in the X–H . . . Y system, but it has gradually been recognized that most other non-metallic atoms can be involved in H-bonds, especially in combination with more electronegative atoms, X–H . . . O or X . . . H–O. Hydrogen bonds involving Cl, Br, I, S, and P have been recognized, and presumably occur with Se and As (not reported). The symmetrical $[HCl_2]^-$ and $[HBr_2]^-$ ions correspond to $[HF_2]^-$ but are much less stable. Thus, the bifluoride ion is formed even with cations as small as Na^+, and it is almost impossible to remove the extra HF from such salts as $CsHF_2$. The bichloride and bibromide are only stable with large cations (e.g. $[Me_4N]^+$) and readily lose HCl or HBr (cf. section 4.9.1). The case of carbon is particularly interesting. A detailed study has been made of the contacts formed by the hydrogen atoms of C–H bonds in 113 organic compounds whose crystal structures have been determined by neutron diffraction (so that accurate hydrogen atomic positions are known) (Taylor & Kennard 1982). Many of the H . . . O distances are more than 30 pm shorter than the expected van der Waals distance (270 pm). This is not the case for other C–H . . . X contacts, which cluster around the van der Waals distances. After eliminating other explanations (such as an incorrect value of the van der Waals radius for oxygen), it became clear that these are genuine C–H . . . O hydrogen bonds. They are formed especially when the C–H bond adjoins an electron-withdrawing group, which produces a fractional positive charge on the hydrogen. A typical example is the C–H . . . O bond formed by 4-hydroxy-proline (I).

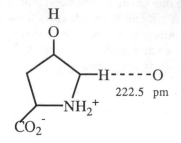

I — 4-hydroxy-l-proline.

7.4.1 The strength of hydrogen bonds

Experimental measurements have been made of the association energies in a great number of hydrogen-bonded systems. Most have involved fairly weak O–H . . . N and O–H . . . O interactions, which can be treated as examples of donor-acceptor bonds. Indeed, Drago's E/C approach (section 7.3.3.2) gives excellent agreement between experiment and calculation with appropriate E_a, C_a values for the O–H group as the acceptor. Typical H-bond energies are those for PhOH . . . O(Et)$_2$ ($25 \, kJ \, mol^{-1}$) and MeOH . . . HOMe ($19 \, kJ \, mol^{-1}$); the latter value relates to one pair of molecules in the gas phase; in liquid methanol a network of H-bonds is formed. The characteristic dimers of carboxylic acids (II) have dimerization energies of about $30 \, kJ \, mol^{-1}$, corresponding of course to the formation of two bonds. The C–H . . . O bonds discussed above must be much weaker still, a few $kJ \, mol^{-1}$ only, comparable to the van der Waals bonds with which they compete in the intermolecular bonding of crystals.

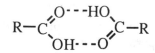

II — Carboxylic acid dimer.

By contrast, in very short hydrogen bonds the interactions are apparently as strong as covalent bonds, though their energies are difficult to determine accurately and generally have to be obtained by theoretical calculations. Examples are H–O–H . . . F$^-$ $98 \, kJ \, mol^{-1}$; $[H_3O]^+$ . . . OH_2 (giving $[H_5O_2]^+$) $151 \, kJ \, mol^{-1}$, and $[F–H–F]^-$, calculated as $220 \, kJ \, mol^{-1}$. The last system has long been considered the strongest hydrogen bond known. However the same type of calculation indicates that the formic acid/F$^-$ bond (III) is even stronger, $250 \, kJ \, mol^{-1}$. This unexpected result needs confirmation by an experimental measurement before we can be fully confident of its correctness.

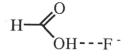

III — Formic acid/fluoride ion hydrogen-bonded system.

7.4.2 The nature of the hydrogen bond

Early descriptions of hydrogen bonding attributed it simply to an electrostatic interaction (corresponding to the dipole-dipole component of van der Waals bonding)

$$X^{\delta-} - H^{\delta+} \ldots Y^{\delta-}$$

The electronegativity of the atoms X and Y makes this a reasonable suggestion, and it also agrees with the strengthening of the bond when X carries a negative charge (as in the $[HX_2]^-$ salts). However it does not explain the linearity of hydrogen bonds or why this becomes more pronounced as the bond strengthens. The typical H ... O–X angle of about 120° (Fig. 7.7) suggests that a lone pair on the second oxygen is involved in the bond; a purely electrostatic interaction would not be expected to show a preference for any particular angle. This additional evidence can best be explained by a covalent or donor-acceptor interaction.

Such an interaction can be viewed in two ways, either as the perturbation of an X–H bond (e.g. O–H) by donation from the approaching lone pair on atom Y (e.g. O), or by the formation of a bond between all three atoms starting from unbonded X, H, and Y atoms. These descriptions are equivalent but the first is more appropriate for weakly bound systems in which the X–H unit retains its identity, the second for strong H-bonds, e.g. [F–H–F]$^-$.

From the donor-acceptor viewpoint (Fig. 7.10), the lone pair on O donates to the σ_{ab} orbital of the O–H bond, creating bonding and anti-bonding molecular orbitals. As there is only a small overlap, the gain in energy will be modest. This donation of electron density into the σ_{ab} orbital will cause the O–H bond to lengthen, to an extent depending on the overlap with the σ_{ab} orbital. This lengthening cannot always be detected in individual crystal structures because the O–H distances are sensitive to

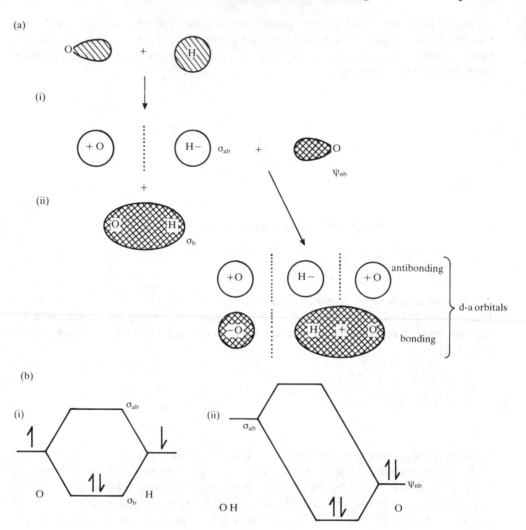

Fig. 7.10 — Donor-acceptor description of hydrogen bond: orbitals and energy diagram. (i) Formation of O–H bond, (ii) overlap with lone pair.

their environment. However, a plot of O–H distance against O . . . O distance (Fig. 7.11) shows a clear correlation; the O–H bond lengthens as the other O approaches (though the individual points have a large scatter about the mean curve). This interaction also explains the decrease in the O–H frequencies in the infrared spectra of H-bonded systems (which is not expected on the electrostatic model). It must be realized that the existence of a donor-acceptor interaction does not exclude electrostatic attraction, though it is difficult to separate the two components. In weak

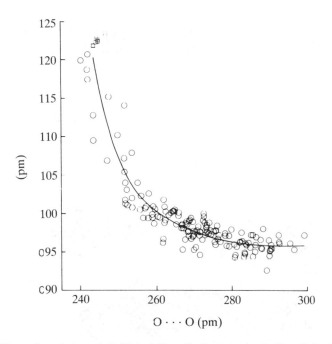

Fig. 7.11 — Correlation of O–H and O...O distances in O–H...O hydrogen bonds.
(Reproduced with permission from Olofsson, I., Jönsson, P. G., In: Schuster, P., Zundel, G.,
Sandorfy, C. (eds) (1976) *The hydrogen bond*, Vol. II, Ch. 8, North-Holland, Amsterdam.)

hydrogen bonds, much of the bond energy is probably derived from electrostatic
forces.

In the bonding description (Fig. 7.12), applied for example to [FHF]$^-$, the three
atomic orbitals (2p or an sp hybrid on each F, 1s on H) overlap to give three
molecular orbitals, bonding, non-bonding and anti-bonding. The system contains
four valence electrons which occupy the σ_b and σ_{nb} orbitals, giving a total bond order
of 0.5 (one bonding pair for two bonds). It is therefore an example of a three-centre/
four-electron system, from the number of atoms and electrons involved (cf. section
5.9.2). This description is particularly appropriate for a symmetrical system, as
observed for [F–H–F]$^-$; as expected for a three-centre/four-electron model, the F–H
distances (113 pm) are longer than in free HF (91.7 pm). The O–H distances in the
symmetrical O–H–O bonds shown in Fig. 7.8 (about 120 pm) are also longer than in
free O–H (100–105 pm).

The importance of the hydrogen bond in chemistry and biology cannot be
underestimated. Among organic compounds, N–H and O–H groups form hydrogen
bonds in almost all molecules of biological importance. Water, the universal
biological medium, is, of course, both a hydrogen bond donor and an acceptor.
These interactions dominate the structures of proteins. Those parts of polypeptide
chains that lack H-bond donors and acceptors are *hydrophobic* (water-hating). The
protein chains are arranged so that these sections are folded inwards on each other

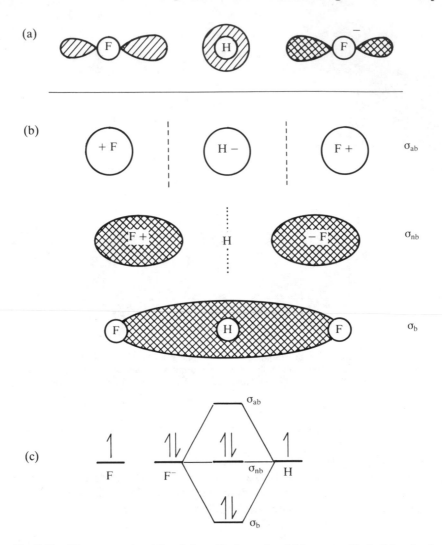

Fig. 7.12 — Three-centre bond description of hydrogen bond (a) parent orbitals, (b) molecular orbitals, (c) energy diagram.

and are held together by van der Waals' interactions. Other parts of the chains are *hydrophilic*, forming strong bonds to water. They will tend to be on the outside of the protein, surrounded by a sheath of water molecules.

Other hydrogen-bond interactions, specific rather than general, play the key role in the genetic code. They bind the pairs of nucleotides together, thymine to adenine (IV) and cytosine to guanine (V), that make the two chains of DNA complementary. Other H-bonds give the DNA chains their helical structure.

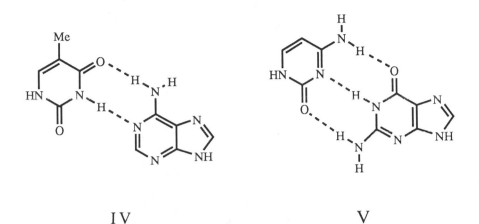

IV   V

IV — Thymine-adenine hydrogen-bonded nucleotide pair.
V — Cytosine-quanine hydrogen-bonded nucleotide pair.

7.5   SECONDARY BONDING

This is one of a variety of terms that are used to describe weak bonds between heavy non-metal atoms, longer than single covalent bonds but shorter than van der Waals' interactions. Secondary bonding has much in common with H-bonding, in its occurrence as an accompaniment to covalent bonding, in its recognition from crystal structure evidence, and in its explanation through dative bonding to σ_{ab} orbitals. It is as important in relation to the heavy non-metals as H-bonding is to N, O, F, etc., but of course is less significant for organic and biological chemistry, simply because these non-metals are less common in that area of chemistry.

A characteristic example illustrating secondary bonding is the IO_3^- ion. Following the VSEPR rules (section 5.6), this is pyramidal with the I–O distance about 180 pm. However, in virtually all iodates the iodine has three other oxygen neighbours completing a distorted octahedron, with I . . . O about 280 pm (Fig. 7.13a), and with O–I . . . O angles close to 180°. These longer distances are the *secondary bonds*. A related compound of particular interest is the acid HIO_3 which contains the same IO_6 structural unit. Normally, oxy-acid crystal structures are controlled by H-bonding, and indeed the crystals of HIO_3 contain H-bonds. However, it has been found that the crystals of HIO_3 have the same structure as those of XeO_3. Clearly the latter contains no H-bonds. It therefore follows that in both compounds the packing

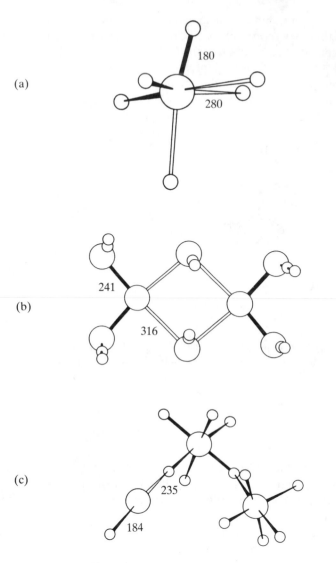

Fig. 7.13 — Secondary bond geometrics (a) octahedral in the iodate ion in $[NH_4]^+[IO_3]^-$, (b) square planar in $[Se_2(SeCN)_6]^{2-}$, (c) linear in $[XeF]^+[Sb_2F_{11}]^-$. Reproduced with permission from Alcock, N. W. (1972) *Adv. Inorg. Chem. Radiochem*, **15** 1.

arrangement depends on the system of secondary bonds, rather than on the hydrogen bonds.

In other compounds, secondary bonds produce square-planar or linear geometries (Fig. 7.13b,c). A particular frequent linear system is the C–Br . . . O unit. This is found in virtually all bromo-organic compounds that contain oxygen atoms. The Br . . . O distances are relatively long (290–320 pm, hardly shorter than the van der

Waals distance of 337 pm), but the linear arrangement occurs so uniformly that there can be no doubt that they represent specific bonds. Comparison with structures with potential hydrogen bonds indicates that the C–Br . . . O bonds are probably similar in strength to weak O–H . . . O bonds. Unfortunately, virtually no other evidence has been found for the strength of secondary bonds.

Some secondary bonded systems have complex geometries, especially when atoms with large polarity differences are involved. These have been examined in detail for $Sb(hal)_3$ molecules, and an example is shown in Fig. 7.14.

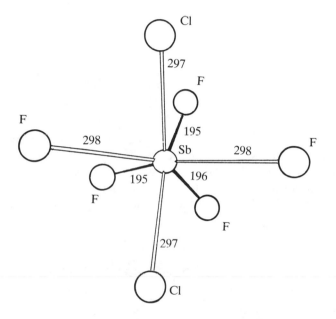

Fig. 7.14 — A complex secondary-bonded structure containing $SbF_3:CsSbClF_3$. Redrawn after Sawyer, J. F. & Gillespie, R. J. (1986) *Prog. Inorg. Chem.* **34** 65.

Secondary bonds can be explained in exactly the same way as hydrogen bonds, by a combination of electrostatic interactions and dative bonding. Like hydrogen bonds, they therefore represent one facet of donor-acceptor behaviour; they are also closely related to some of the charge-transfer complexes describe above (e.g. Fig. 7.4). However, secondary bonds are principally examined and identified in solids, charge-transfer complexes in liquids, gases, or solutions. Occasionally, solids give a good impression of liquid structures, and an interesting example of this is the complicated structure of $2[NMe_4]^+[(SeOCl_2)_{10}Cl_2]^{2-}$ (Fig. 7.15). The molecules of $SeOCl_2$ are entirely held together by Se–O . . . O and Se–O . . . Cl secondary bonds, and this confused arrangement must be very similar to that in liquid $SeOCl_2$.

In a linear secondary bond, the acceptor orbital is presumed to the σ_{ab} orbital, exactly as for H-bonding. Fig. 7.12 is therefore a precise description of the molecular orbitals in an O–I . . . O secondary bond if I is substituted for H. Most linear

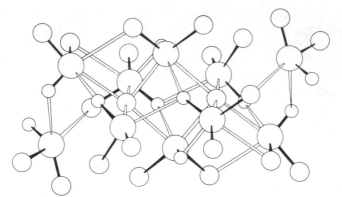

Fig. 7.15 — Structure of $2[NMe_4]^+[(SeOCl_2)_{10}Cl_2]^{2-}$ showing the complex network of
secondary bonds. Source: as Fig. 7.13.

secondary X–Y...Z systems, like most H-bonds, are asymmetrical, but in some
examples the X–Y bond has a wide range in length, and the Y...Z distance shortens
as X–Y lengthens (Burgi 1975). These examples include Cl–Sb...Cl interactions,
and S...S...S bonds in thiathiopthenes (VI), in which the distances vary as the R-
substituents change. The I...I...I unit is particularly notable. Molecules of I_2

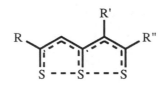

VI — A thiathiophthene.

themselves interact weakly in the solid, and the I–I distance is detectably lengthened
(270 pm, compared to 267 pm for I_2 in the gas phase). This trend can be followed
through a great range of poly-iodide ions, such as Cs_2I_8, in which I_2 is weakly bonded
to two I_3^- units, or I_3^- itself, which can be symmetrical or unsymmetrical.

These bond lengths correspond very closely to the predictions for a three-centre
system in which the individual bond orders n_1 and n_2 give a total bond order of 1.0,
i.e.

$$n_1+n_2=1.0 \tag{7.5}$$

and each bond order is related to bond length by the relationship

$$\log_{10}n=(r_0-r)/c \tag{7.6}$$

(section 5.8), with r_0 and c being empirical constants (2.67 pm and 85 pm for this example). The solid line in Fig. 7.16 shows the excellent fit of the observed

Distances in I---I---I units (pm)

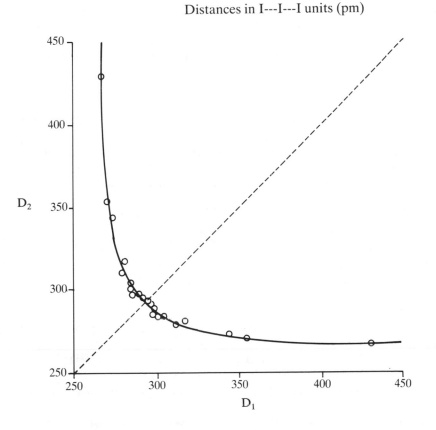

Fig. 7.16 — Correlation of I---I---I distances in $[I_3]$ units. The solid line corresponds to the formula given in the text. Each point is plotted twice ($D_1 > D_2$ and $D_2 > D_1$), reflected in the 45° line. Source: redrawn from Burgi, H.-B. (1975) *Angew. Chem. Int.* **14** 460.

I---I---I distances to the hyperbolic curve, given by equation (7.7)

$$r_2 = r_0 - c(\log_{10}\{1.0 - 10^{[(r_0 - r_1)/c]}\}) \tag{7.7}$$

The I..I..I system therefore gives an excellent view of the correlated variation of bond lengths in three-centre/four-electron systems. This curve also corresponds exactly to the behaviour of O–H...O systems, except that the latter are almost always unsymmetrical, with one strong and one weak interaction.

FURTHER READING

Van der Waals bonding

Bondi, A. (1964) *J. Phys. Chem.* **68**, 441.

Day, M. C. & Selbin, J. (1969) *Theoretical inorganic chemistry*, Reinhold, New York.

Maitland, G. C., Rigby, M., Smith, E. B. & Wakeham, W. A. (1981) *Intermolecular forces, their origin and determination*. Clarendon Press, Oxford.

Donor-acceptor bonding

Drago, R. S. (1973) Quantitative evaluation and prediction of donor-acceptor interactions. *Structure and Bonding* **15** 73.

Pearson, R. G. (1963) Hard and soft acids and bases. *J. Amer. Chem. Soc.* **85**, 3533.

Pearson, R. G. (1973) *Hard and soft acids and bases*. Dowden Hutchinson Ross, Strousberg, Pa. (Reprints with commentary a number of important papers, including the preceding one.)

Hydrogen-bonding

Emsley, J. (1980) Very strong hydrogen-bonding. *Chem. Soc. Rev.* **9** 91.

Schuster, P., Zundel, G., & Sandorfy, C. (eds) (1976) *The hydrogen bond*, North-Holland, Amsterdam.

Secondary bonding

Alcock, N. W. (1972) Secondary bonding to non-metallic elements. *Adv. Inorg. Chem. Radiochem.* **15** 1.

Sawyer, J. F. & Gillespie, R. J. (1986) The stereochemistry of Sb(III) halides and some related compounds. *Prog. Inorg. Chem.* **34** 65.

Burgi, H.-B. (1975) Stereochemistry of reaction paths. *Angew. Chem. Internat.* **14** 460.

Text references

Burkert, U. & Allinger, N. L. (1982) *Molecular mechanics*. American Chemical Soc., Michigan. Monograph 177, p. 310.

Taylor, R. & O. Kennard (1982) *J. Amer. Chem. Soc.* **104** 5063.

8

Dative bonding in transition metal complexes

8.1 INTRODUCTION

Transition metal complexes are just one group within the broad area of donor-acceptor complexes, but the involvement of the d orbitals on the metal ions makes their bonding more complex than the examples already considered. For this reason, as well as their general importance, they are described separately. However, they are discussed in detail in a number of textbooks, and this chapter therefore considers only the most important type, the octahedral metal complexes. In these, the metal ion is surrounded by six donor atoms or groups (the *ligands*, so called because they are 'bound' to the metal) at the corners of an octahedron — which can also be described, of course, as the + and − ends of the x, y and z axes. The reader is referred to the texts listed at the end of this chapter for fuller details, especially for the bonding and properties of complexes with other geometries. The bonding in complexes of non-transition metals is essentially identical, but they do not show the effects due to d orbitals.

Transition metal complexes can be approached from two points of view. The simplest is to treat them as ionic. This is a good approximation for only a very few compounds, such as MnF_2 or $[MnF_6]^{4-}$, but it does give useful information about the behaviour of the d electrons in an octahedral environment. A more realistic treatment for most complexes examines the dative interaction between the lone pair orbitals on the donor atoms and empty orbitals on the metal. Conveniently, the qualitative predictions of this model are the same as those obtained from the ionic viewpoint. The properties of transition metal complexes, such as their spectroscopic and magnetic behaviour, can therefore be studied without needing to decide which bonding model is the more appropriate.

8.2 IONIC COMPLEXES

An isolated Mn^{2+} ion contains five d electrons, one in each d orbital. In the gas phase, the five orbitals are *degenerate* (equal in energy). However, they cease to be so if the Mn^{2+} ion is approached by six F^- ions at the corners of an octahedron (the geometry found in solid MnF_2 and in the $[MnF_6]^{4-}$ ion). The Mn^{2+} cation and the F^- ions are attracted by electrostatic forces, but the individual negatively-charged electrons experience a repulsion from the negatively-charged fluoride ions. As these ions get closer, the electrons in different orbitals will be affected differently. The lobes of the $d_{x^2-y^2}$ orbital point directly at four F^- ions in the $x-y$ plane (Fig. 8.1a).

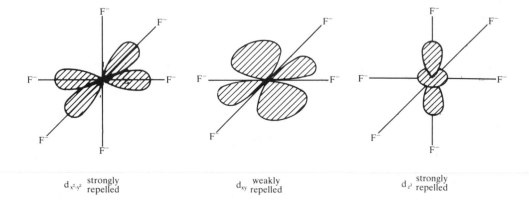

$$d_{x^2-y^2} \quad \text{strongly repelled} \qquad\qquad d_{xy} \quad \text{weakly repelled} \qquad\qquad d_{z^2} \quad \text{strongly repelled}$$

Fig. 8.1 — The repulsive effect of six F^- ions on electrons in the individual d-orbitals of Mn^{2+} (a) $d_{x^2-y^2}$: strongly repelled, (b) d_{xy}: weakly repelled, (c) d_{z^2}: strongly repelled.

The d_{xy} orbital is the same shape as the $d_{x^2-y^2}$ orbital but rotated through 45°. Its lobes therefore point *between* the four F^- ions, and an electron in this orbital is less strongly repelled than one in the $d_{x^2-y^2}$ orbital (Fig. 8.1b). The d_{yz} and d_{xz} orbitals are identical to the d_{xy} orbital, apart from their different orientations, and so are also moderately affected. However, the d_{z^2} orbital, like $d_{x^2-y^2}$, has lobes pointing along the z axis which are strongly influenced by the approaching F^- ions; it can be shown that d_{z^2} remains degenerate with $d_{x^2-y^2}$ (Fig. 8.1c). The overall effect on the orbital energy is shown in Fig. 8.2. Taken as a whole, the approaching ions raise the d orbital energies, but individually the orbitals split into two groups, three lower in energy (less strongly affected) and two higher. These groups are conventionally labelled t_{2g} and e_g, and the gap between them is labelled Δ.†

This approach to transition metal complexes is called *crystal field theory*, as it describes the effect of the octahedral electrostatic field found in crystalline MnF_2. The splitting between the two groups of orbitals, Δ, is called the *crystal field splitting*. The effects of this splitting are the main concern of this section.

8.2.1 The spectrochemical series

The separation of the d orbitals into the t_{2g} and e_g groups can be studied spectroscopically. An electron in one of the t_{2g} orbitals can be promoted to one of the e_g orbitals by the absorption of a photon with a frequency in or near the visible range (Fig. 8.3).

† Sometimes the symbol 10Dq is used instead of Δ.

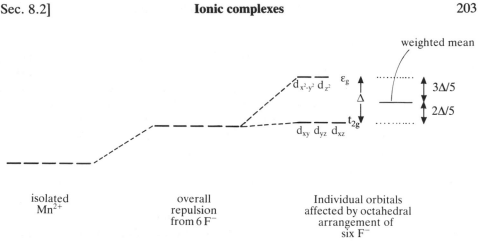

Fig. 8.2 — Overall effect of approaching F^- ions on d orbital energies.

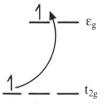

Fig. 8.3 — Absorption of light by an octahedral metal complex with one d-electron.

This allows Δ to be determined. Its values depend both on the type of donor group and on the metal. Thus, different transition metals complexes have characteristic colours, and changing the ligand changes the colour, e.g. pale blue for $Cu(H_2O)_4^{2+}$, deep blue for $Cu(NH_3)_4^{2+}$, yellow for $CuCl_4^{2-}$. If the ligands are arranged in the order of the Δ values that they produce, almost the same order is found for different metals. This is known as the *spectrochemical series*, because it correlates spectroscopic behaviour with chemistry. For common ligands the order is:

$$I^- < Br^- < Cl^- < F^- < OH^- < acetate < H_2O < pyridine \approx NH_3 < CN^- < CO.$$

This in no way corresponds to what would be expected if the influence of the ligands on the d orbitals was purely electrostatic; neutral ligands like NH_3 or CO would be expected to have very little effect, rather than producing much larger d orbital splittings than small charged ligands like F^-. This major flaw in the crystal field approach is rectified when covalent bonding between metal and ligand is considered (Section 8.3).

8.2.2 Crystal field stabilisation energy

Consider what happens as the number of d electrons is increased in an octahedral MF_6 complex. In a d^1 complex, the electron naturally occupies one of the lower (t_{2g}) orbitals (Fig. 8.4). The complex of course has one unpaired electron and is paramagnetic (Box 5). This electron is somewhat more stable than the mean energy of the d orbitals (indicated by the line in Fig. 8.2). This additional stabilization depends on the size of the crystal field splitting, Δ, and for this single electron is given by $2\Delta/5$. This is known as the *crystal field stabilization energy* (CFSE).†

In d^2 and d^3 complexes, the next electrons will occupy the other t_{2g} orbitals, parallel to the first electron (following Hund's rule). They have respectively CFSE of $4\Delta/5$ and $6\Delta/5$, and 2 or 3 unpaired electrons. With d^4 we encounter a problem. This fourth electron has two unsatisfactory possibilities open to it. It can either enter an e_g orbital of relatively high energy, or it can pair with one of the electrons already in the t_{2g} orbitals, leading to increased electron–electron repulsion and destabilization by the pairing energy. The choice depends on the magnitude of Δ. If this exceeds the pairing energy, then the electron will enter a t_{2g} orbital; if not, it will go into an e_g orbital.

These alternatives can be distinguished by the paramagnetism of the complex. This corresponds either to 2 unpaired electrons (large Δ) or to 4 unpaired electrons (small Δ). The two possibilities are therefore labelled respectively *low spin* and *high spin* complexes. From d^6 onwards, even the high spin complexes need to have some electrons paired, while in a low spin d^7 complex one electron occupies an e_g orbital. Finally, from d^8 to d^{10}, the t_{2g} orbitals are full. High and low spin complexes are no longer distinguished.

Fig. 8.5 shows the overall pattern of CFSE for high spin complexes, and Fig. 8.6 illustrates two real examples: the lattice energies (experimental values) of MCl_2 compounds and the hydration energies of the M^{2+} ions of the first transition series. The dashed lines show the smooth curves through the points for Ca^{2+}, Mn^{2+} and Zn^{2+}. These ions are chosen because they have zero CFSE. The double cusped form of the experimental curves corresponds very well to what is expected from the additional stabilization provided by the crystal field splitting.‡ The similarity of the curves indicates that values of Δ for a given ligand do not vary much as the metal changes, (though for M^{3+} they are about 1.5 times the M^{2+} values). The Δ values for six H_2O ligands are known for all the first row ions, so that the hydration energies can be corrected to CFSE-free values. These values (circles in Fig. 8.6b) lie close to the smooth line through Ca^{2+}, Mn^{2+}, and Zn^{2+}. It is worth noting that the largest CFSE values are quite a small proportion (about 10%) of the total lattice or hydration energy.

One complication has been ignored in this description, the repulsive energy needed to pair two electrons (the pairing energy). This has to be supplied in high spin complexes when six or more d electrons are present. However, in the lattice energy calculations it cancels out because this energy relates to the process

$$MCl_2 \, (s) {\rightarrow} M^{2+} \, (g) + 2 \, Cl^-.$$

---

† The reason for measuring these energies relative to the mean energy of the d orbitals will appear shortly.
‡ The same double-cusped curve, characteristic of crystal–field stabilization, can be seen in the curve showing the metallic bond energy of the transition metals (Fig. 6.12).

d^1 – d^3

	d$_1$	d$_2$	d$_3$
CFSE =	$-2\Delta/5$	$-4\Delta/5$	$-6\Delta/5$
unpaired electrons	1	2	3

d^4 – d^7

High spin (small Δ)

	d^4	d^5	d^6	d^7
CFSE =	$-3\Delta/5$	0	$-2\Delta/5$	$-4\Delta/5$
unpaired electrons	4	5	4	3

d^4 – d^7

Low spin (large Δ)

	d^4	d^5	d^6	d^7
CFSE =	$-8\Delta/5$	$-10\Delta/5$	$-12\Delta/5$	$-9\Delta/5$
unpaired electrons	2	1	0	1

d^8 – d^{10}

	d^8	d^9	d^{10}
CFSE =	$-6\Delta/5$	$-3\Delta/5$	0
unpaired electrons	2	1	0

Fig. 8.4 — Addition of d-electrons to an octahedral complex: crystal field stabilisation energy (CFSE) and paramagnetism.

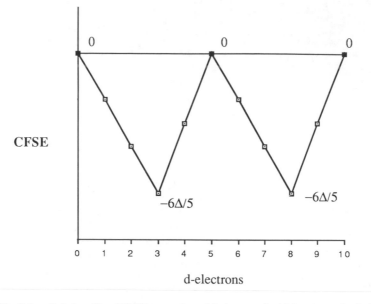

Fig. 8.5 — Relationship of CFSE to number of d-electrons for high-spin octahedral complexes.

The M^{2+} (g) is the isolated gas-phase ion with all its d orbitals of equal energy. For d^6 as an example, this has two electrons paired, so the pairing energy is unchanged on forming a high spin crystal. However, in a crystal containing a low spin d^6 ion, four more electrons are paired up and the experimental lattice energy will be reduced by two additional units of pairing energy, as well as including the CFSE of $-12\Delta/5$.

8.2.3 Ionic radii and crystal field effects
The electron repulsion that produces crystal field effects also operates in reverse. As the e_g orbitals point directly at the ligands, electrons in these orbitals physically repel the ligands more than do electrons in the t_{2g} orbitals. Thus, ions with electrons in t_{2g} orbitals are smaller than those with electrons in e_g orbitals. The effect of this can be seen in Fig. 8.7. As with Fig. 8.6, the smooth curve shows what happens in the absence of crystal field effects: a smooth decrease as the nuclear charge increases. For high spin ions, the radii fall slightly below these values as the first electrons enter the t_{2g} orbitals; they return to the smooth curve at d^5 and then repeat the behaviour. Low spin ions (d^4-d^7) have as many electrons as possible in the t_{2g} orbitals and so are considerably smaller than the corresponding high spin ions. This has a number of important structural consequences. One concerns the binding of O_2 by iron in haemoglobin (Fig. 8.8). In the free protein, the ion is five-coordinate high spin Fe^{2+} (d^6), bound to the four N-atoms of a porphyrin ligand and to a histidine amino acid. It is too large to fit into the hole in the ligand and therefore sits on one side of it. When it binds O_2, it becomes low spin and will fit into the ring. The effect of the metal ion

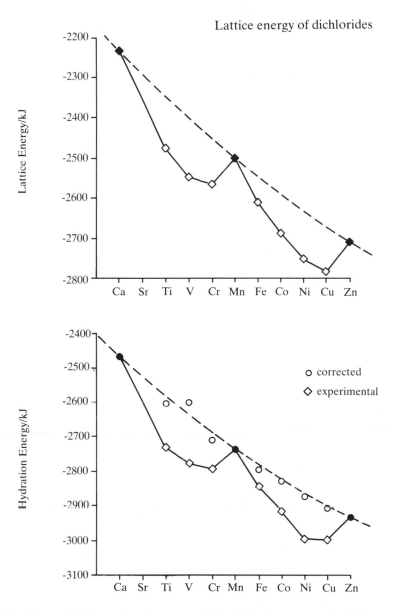

Fig. 8.6 — Experimental evidence for CFSE (a) Lattice energies for MCl$_2$ compounds (experimental values from Born-Haber cycle calculations), (b) Heats of hydration of [M(H$_2$O)$_6$]$^{2+}$. Circles indicate the result of deducting the CFSE calculated from spectroscopic values of Δ. Source: data in George, P. & McClure, D. S. (1959) *Progress in Inorg. Chem.* **1** 381.

moving is to pull the histidine ligand about 80 pm towards the ring. This leads to changes in the structure of the rest of the protein, increasing its O$_2$ binding ability.

The radius variations shown in Fig. 8.7 have another important implication. For

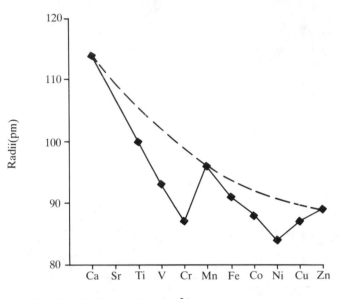

Fig. 8.7 — Radii of octahedral M^{2+} ions. Source: as Appendix A.4.

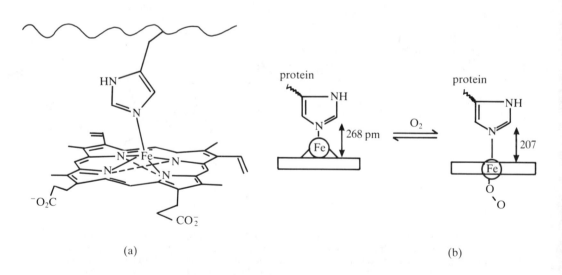

Fig. 8.8 — Binding of O_2 to haemoglobin (a) coordination of Fe by porphyrin and histidine, (b) side view, O_2 absent, (c) side view, after addition of O_2. Reproduced with permission from (a) Cotton, F. A. & Wilkinson, G. (1988) *Advanced inorganic chemistry*, Wiley, New York, 5th ed. p. 1347. (b) Hughes, M. N. (1981) *Inorganic chemistry of biological processes*. Wiley, Chichester, p. 228.

an ionic compound, the lattice energy depends on the interionic distance (roughly as $1/r$; section 4.9) and therefore on the radii of the individual ions. Thus, the reduction in radius shown in Fig. 8.7 itself increases the lattice energy, independent of the electronic CFSE. The effect can be examined for NiF_2, one of the few compounds for which data exist both for the lattice energy (3059 kJ mol^{-1}) and for Δ (87 kJ mol^{-1} for $NiF_6{}^{4-}$). (Stout & Reed 1954, Jørgensen 1962). The radius shrinkage for Ni^{2+} can be estimated either from Fig. 8.7 or from the MF distance in NiF_2 (200 pm), compared to those in MnF_2 and ZnF_2, as about 6 pm. If we assume that for this small decrease in radius, the lattice energy varies as $1/r$, it will be increased by 206/200, i.e. about 90 kJmol^{-1}. The CFSE of $6\Delta/5$ corresponds to 104 kJ mol^{-1}, so in the absence of any crystal field effects the lattice energy would be 2865 kJ mol^{-1}, slightly smaller than the value obtained by interpolation between MnF_2 and ZnF_2 (2902 kJ mol^{-1} with an error of about 20 kJ mol^{-1}).

The effect of the radius change is at its largest with small anions such as F^- or O^{2-}, for which this calculation shows that it is as important as the direct effect of the crystal field stabilization. However, this does leave one problem unresolved. The hydration energies of the $[M(H_2O)_6]^{2+}$ ions should also be *directly* affected by the changing radius of the metal ions, as well as by the CFSE. It is surprising, therefore, that the correction for CFSE alone appears to explain the whole of the observed deviation from a smooth curve. It may be that the relatively weakly bound H_2O ligand is affected less by repulsion from electrons in e_g orbitals than is the small F^- ion.

8.2.4 Distorted geometries: the Jahn–Teller effect

The final consequence of crystal field theory to be examined relates to electronic configurations which have unequally occupied e_g orbitals. Consider the d^9 configuration in Fig. 8.4, with one electron in one e_g orbital and two in the other, and assume for the moment that the pair of electrons is in the d_{z^2} orbital. Figs 8.1a,c show that the two ligands at each end of the z-axis will be more strongly repelled by the pair of electrons in the d_{z^2} orbital than are the four ligands in the $x-y$ plane by the single electron in the $d_{x^2-y^2}$. Such an octahedral complex distorts so that the four ligands in the $x-y$ plane are closer, and the two on the z-axis are more distant. Exactly this arrangement is found in many d^9 Cu^{2+} complexes, such as $[NH_4]_2CuCl_4$, with four shorter Cu-Cl distances (230 pm) and two longer (279 pm) (Fig. 8.9).

This is an example of a general effect, described as a Jahn–Teller distortion. The Jahn–Teller theorem states that *any non-linear structure with a degenerate electronic configuration will distort to remove the degeneracy of the electronic configuration.*[†] For the d^9 configuration, the two degenerate states are $(d_{x^2-y^2})^2(d_{z^2})^1$ and $(d_{x^2-y^2})^1(d_{z^2})^2$. The theorem does not state what form the distortion will take nor how large it will be. Generally, Cu^{2+} complexes have *two* long and *four* short bonds as just described, though in principle, Cu^{2+} complexes might have *two* short bonds and four longer ones, with the single electron in the d_{z^2} orbital. Many textbooks report that K_2CuF_4 shows the latter bond pattern. However, re-examination of the structure has shown that it has the standard pattern of four short and two long Cu−F bonds. No confirmed examples of the $2+4$ pattern are known, though distorted

---

[†] The effect has already been encountered in NO_2 (section 5.9.3).

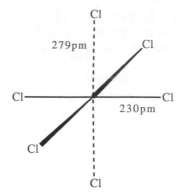

Fig. 8.9 — Coordination of Cu^{2+} in $[NH_4]_2CuCl_4$.

structures of 5-coordinate Cu^{2+} have been discovered. It is also possible for the distortion to be dynamic. In compounds showing dynamic Jahn–Teller distortions, the CuL_6 octahedron is regular on average, but instantaneously the distances to one of the three pairs of ligands is longer than the other distances.

Jahn–Teller distortions are not limited to d^9 configurations. Consideration of Fig. 8.4 shows that they will also be found for d^4 high spin, e.g. Cr^{2+}, and d^7 low spin, e.g. Co^{2+} or Ni^{3+}. The former is common, and many examples are known of distorted Cr^{2+} complexes, but low spin Co^{2+} and Ni^{3+} are virtually unknown. In principle, the Jahn–Teller theorem must apply to many configurations with degenerate t_{2g} configurations (d^1, d^2, etc.). However, no distorted structures have been observed for complexes with these configurations, and it is believed that the distortions are all dynamic. Also, they are probably small, because the t_{2g} orbitals point *between* rather than at the ligands.

8.3 MOLECULAR ORBITAL TREATMENT OF METAL COMPLEXES

The preceding section shows that crystal field theory gives an excellent explanation for the properties of transition metal complexes. This is despite the implausibility of its original postulate of purely electrostatic ligand–metal interaction. The much more reasonable molecular orbital approach treats metal complexes as containing covalent donor–acceptor bonds. Most important, *it predicts the same d orbital splitting* as does crystal field theory. It is therefore equally good at explaining the properties of metal complexes. This molecular orbital approach is often named as *ligand field theory*, in parallel with crystal field theory. The various crystal field parameters (the splitting Δ, CFSE, and others not discussed here) are then called '*ligand field splitting*', etc. It seems more satisfactory to retain the names these parameters were originally given, and to avoid the meaningless term 'Ligand Field', while recognizing that the molecular orbital treatment gives a better explanation of the origin of crystal field splitting than does the original theory.

8.3.1   σ molecular orbitals

Applying molecular orbital theory to complexes is made difficult by the variety of possible ligands, each with their individual molecular orbitals available for interaction with the metal orbitals.†

Fortunately, almost all ligands have one common feature that underlies the molecular orbitals for the major component of ligand–metal bonding. Each has a lone pair in an orbital directed towards the metal — the lone pair that gives it donor character. The rare exceptions are ligands which donate from bonding σ orbitals (class 5 in Table 7.2), e.g. C_2H_4 in $K+[PtCl_3C_2H_4]^-$ (Zeise's salt). These are beyond the scope of this chapter.

The symmetry of these ligand orbitals in octahedral complexes shows that all are of σ character, because the lobe of the lone pair points *along the ligand-metal direction*. Ligands of π character are discussed later. The ligand σ orbitals are shown in Fig. 8.10a. They interact with orbitals on the metal that are also of σ symmetry, i.e.

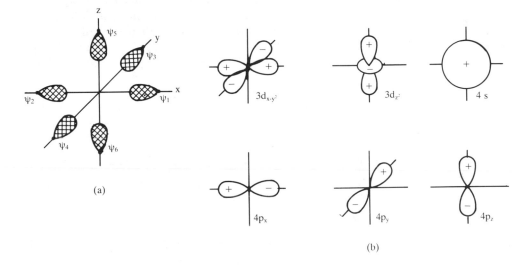

(a)

$3d_{x-y^2}$ $3d_{z^2}$ 4 s

$4p_x$ $4p_y$ $4p_z$

(b)

Fig. 8.10 — Orbitals of σ-symmetry (a) ligand orbitals, (b) metal orbitals.

also have lobes which point along the ligand-metal directions. For first-row transition metals, these comprise the $3d_{x^2-y^2}$, $3d_{z^2}$, 4s, and $4p_x$, $4p_y$, $4p_z$ orbitals (Fig. 8.10b); the remaining metal d orbitals (d_{xy}, d_{xz}, d_{yz}) are of π symmetry. It might be supposed that the p orbitals should be of π symmetry (as in ethene) because, for example, the $4p_y$ and $4p_z$ orbitals have a change of phase across the line between the metal and the ligand on the x-axis. However, they have no phase change in the y-direction and are

---

† It is possible to take all the ligand and metal atomic orbitals and build up the total molecular orbitals of the metal complex from these. However, as the ligand molecules exist independently and have structures that are not greatly altered by coordination, it is easier to treat the ligand–metal bonding starting from the ligand molecular orbitals.

therefore σ orbitals. As the next section shows, metal π orbitals have sign changes when viewed from *any* bond direction. Note particularly that the metal 4s and 4p orbitals are used. Even though they lie rather higher in energy than the d orbitals, their overlap with the ligand orbitals is strong enough for significant bonding.

Following the simple approach used for covalent compounds (section 5.4.1), we would now form six d^2sp^3 hybrid orbitals and then produce sets of six equivalent bonding and six equivalent anti-bonding orbitals, by allowing each hybrid orbital to interact with one ligand orbital. This is an oversimplification for metal complexes, and it is necessary to take a more rigorous view in order to establish the energy pattern for the σ orbitals. The ligand orbitals ($\psi_1-\psi_6$) are combined into sets by using appropriate linear combinations, each of which has the right symmetry to overlap with one of the metal orbitals. These combinations are known as *terminal atom symmetry orbitals* (TASO). Each TASO then interacts with the appropriate metal orbital. (Purcell & Kotz 1977). The TASOs are labelled here according to their symmetry, for consistency with the descriptions in other texts.

The appropriate TASO to overlap with the 4s orbital is obvious: ($\psi_1 + \psi_2 + \psi_3 + \psi_4 + \psi_5 + \psi_6$) (the a_{1g} TASO) (normalisation constant omitted for simplicity). For the $4p_x$ orbital, the combination is ($\psi_1 - \psi_2$), with equivalent sets for $4p_y$, $4p_z$ (the three t_{1u} TASOs). Finally, two TASOs, of equal energy though different appearance, combine with the $3d_{x^2-y^2}$ and $3d_{z^2}$ orbitals (the e_g TASOs) (Figs 8.11a–c).

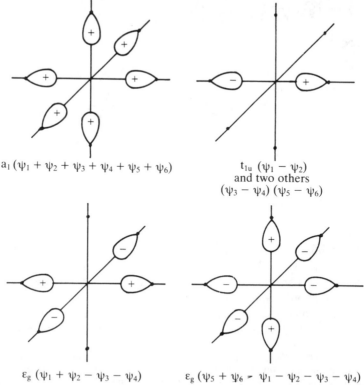

$$a_1 \, (\psi_1 + \psi_2 + \psi_3 + \psi_4 + \psi_5 + \psi_6)$$

$$t_{1u} \, (\psi_1 - \psi_2)$$
and two others
$$(\psi_3 - \psi_4) \, (\psi_5 - \psi_6)$$

$$\varepsilon_g \, (\psi_1 + \psi_2 - \psi_3 - \psi_4)$$

$$\varepsilon_g \, (\psi_5 + \psi_6 \succ \psi_1 - \psi_2 - \psi_3 - \psi_4)$$

Fig. 8.11 — Terminal atom symmetry orbitals (TASOs) for ligand-metal σ-bonding.

Forming molecular orbitals by combining the metal orbitals and the TASOs is now simple (Fig. 8.12). Each pair of orbitals produces a corresponding bonding and anti-bonding molecular orbital. The orbital overlap and hence the splitting between bonding and anti-bonding orbitals is in the order s>p>d. The diagram also shows the d_{xy}, d_{xz} and d_{xy} metal orbitals (the t_{2g} set) which are unchanged in energy, because they do not interact with the σ ligand orbitals. The ligand electrons can now be inserted (shown in Fig. 8.12). Each ligand provides one pair of electrons, its lone pair. These six pairs occupy the bonding a_1, t_{1u}, and e_g molecular orbitals. It is the decrease in energy of these electrons, compared to the free ligand, that provides the bond strength of the complex.

For the metal electrons, consider the part of the diagram within the dotted box, comprising the t_{2g} metal π orbitals and the e_g* orbitals. *This shows exactly the same splitting pattern as is found using crystal field theory.* We can therefore label the splitting between t_{2g} and e_g* as Δ. The electron arrangement depends on the number of d electrons and the value of Δ, exactly as shown in Fig. 8.4. The molecular orbital approach gives a much more plausible explanation of how the splitting between the t_{2g} and e_g* orbitals arises, but is in complete agreement with crystal field theory in its predictions of the properties of complexes.

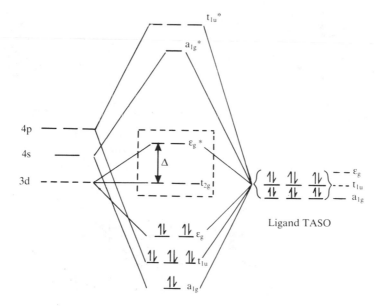

Fig. 8.12 — Molecular orbital energy diagram for σ-bonding in a ML_6 metal complex, showing the six ligand electron pairs. The dotted box indicates the set of five orbitals that contain the metal d-orbitals. Antibonding orbitals are starred.

8.3.2 π-Molecular orbitals

By examining the π orbitals in metal complexes, we can also use the molecular orbital approach to understand the ordering of the spectrochemical series. On the metal atom, the d_{xy}, d_{xz}, and d_{yz} orbitals are of π symmetry, changing sign across the metal-ligand bond axis (Fig. 8.13a). A variety of ligand orbitals of π symmetry exist, but they fall into two distinct classes:

(a) low energy and therefore *filled*. The most important examples are lone-pair p orbitals, e.g. on F⁻ (oriented perpendicular to the bond axis; Fig. 8.13b)
(b) high energy and therefore *empty*. (i) empty d orbitals, e.g. on P in a phosphine ligand, PR_3 (Fig. 8.13c). (ii) π_{ab} orbitals of a π bonded ligand, e.g. CO or CN⁻ (Fig. 8.13d). The bonding π orbitals of these ligands in principle belong to class (a), but are concentrated on the ligand and do not interact strongly with the metal orbitals.

Some ligands, such as Cl⁻, possess both types of orbital, but their interactions are usually dominated by one type. Metal-Cl bonding normally involves the filled p orbitals rather than the empty d orbitals.

The effect of interactions with the π orbitals on the metal can be described by treating them as perturbations of the main molecular orbital scheme (Fig. 8.12). The

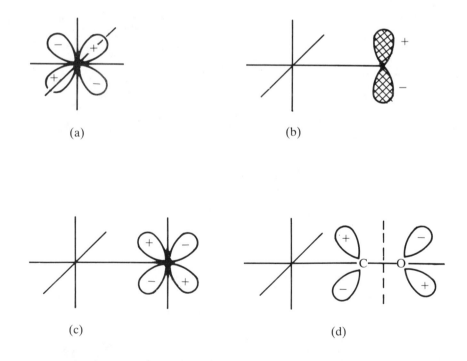

(a)   (b)

(c)   (d)

Fig. 8.13 — Metal and ligand orbitals of π-symmetry (a) metal d_{xz} orbital (d_{xy} and d_{yz} are similar), (b) ligand p-orbital of π-symmetry, (c) empty ligand d-orbital (e.g. d_{xz}), (d) ligand π* orbital.

π molecular orbitals will be similar to the parent orbitals, with some splitting in energy caused by the interaction. Whichever orbital was originally lower in energy will be slightly lowered by the interaction, while the higher energy orbital will be raised.

8.3.2.1   *Low-energy filled π orbitals*

The first type of interaction is between the metal π orbitals and ligand orbitals of type (a): *low energy* (and therefore filled). There may be as many as 12 such orbitals (e.g. on six F^- ligands), but this number is not important. Because the interaction is relatively small, the result will be as shown in Fig. 8.14a. The bonding molecular orbitals will be similar to the original ligand π orbitals, slightly lowered in energy,

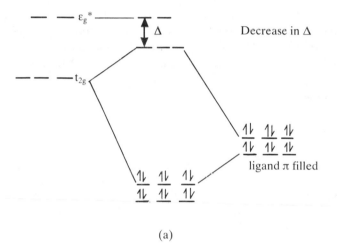

(a)

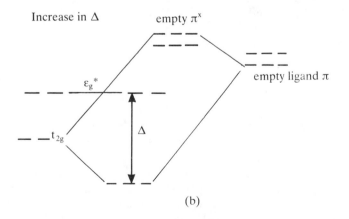

(b)

Fig. 8.14 — Interactions between metal and ligand π orbitals (a) with low-energy filled ligand π orbitals, (b) with high-energy empty ligand π orbitals.

while the anti-bonding π molecular orbitals will resemble the metal t_{2g} orbitals and be slightly raised in energy. The π electrons originally in the ligand orbitals will occupy the π bonding orbitals. As these molecular orbitals now include a small proportion of metal orbital character, the ligand π electrons will move slightly towards the metal. These ligands are therefore called π *donor ligands*. As before, the t_{2g} and e_g will contain the metal d electrons, but the value of Δ will be *reduced*.

8.3.2.2   High-energy empty π orbitals

The reverse interaction occurs with ligands having type (b) orbitals: *high energy* (and therefore empty) (Fig. 8.14b). Now it is the π_{ab} orbitals which are primarily of ligand character, and these remain empty. The π bonding orbitals are primarily of metal t_{2g} type and they are lowered in energy i.e. Δ is *increased*. These orbitals acquire some ligand-orbital character. Electron density will move towards the ligands, which are therefore π *acceptor* ligands. When the ligands are molecules like CO or CN^-, the π bonding molecular orbital has some anti-bonding character in the CO or CN bond, weakening this bond (Fig. 8.15). This can readily be seen in the i.r. stretching frequency of the bond, which provides a sensitive probe for the amount of π overlap (e.g. CO alone, 2143 cm^{-1}, $Co(CO)_6$ 2002 cm^{-1}).

These interactions can be summarised as follows:

Ligand π orbital	Filled low-lying	Empty high energy
Effect on Δ	decrease	increase
Electron donation	ligand $\rightarrow$ metal	metal $\rightarrow$ ligand
Oxidation state stabilization	high oxidation states	low oxidation states
Ligand type	π-donor	π-acceptor

The effect on Δ of the two types of ligand allows us to understand the spectrochemical series (section 8.2.1). π Donor ligands, such as halide ions, produce small splittings and lie at the left end of the series. π Acceptor ligands, such as CO, produce large Δ values and are at the right end of the series.

The effect on oxidation state indicated in the table is crucial to organometallic chemistry. The donation of electrons from metal to π acceptor ligands gives the metal stability in low formal oxidation states, by moving some of its electron density to the ligands. This also has a remarkable effect on the nature of the metal itself, which can best be seen as increasing its effective electronegativity. As a result, metals in low oxidation states form stable covalent bonds to alkyl and aryl groups and H-atoms. For metals in higher oxidation states, such bonds are generally unstable, either thermally or to oxidation or hydrolysis.

A third intriguing consequence of the large Δ produced by π acceptor ligands is the strong tendency to fill only the three t_{2g} orbitals, making the e_g orbitals truly anti-bonding. Bearing in mind that the σ bonding metal-ligand orbitals are formed from six metal orbitals, 4s, 4p, and 3d (e_g) orbitals, a total of 18 electrons are needed to fill all these orbitals. This is the basis for the 18-electron rule, used to predict the stability of low oxidation-state complexes. These complexes need not be octahedral. If they

(a)

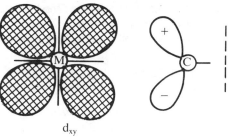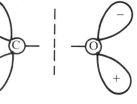

d_{xy} π_{ab}

(b)

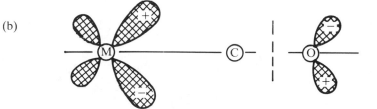

bonding combination

(c)

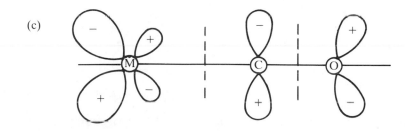

antibonding combination

Fig. 8.15 — π-interactions between a metal and CO (a) orbitals before interaction, (b) bonding molecular orbital, with some anti-bonding character in the C–O bond, (c) antibonding molecular orbital.

have fewer than six ligands, then some of the bonding orbitals become lone pair orbitals, but will still be filled in a stable complex.

FURTHER READING

Most textbooks discuss bonding and properties of transition metals complexes in some detail. The following are suggested as useful sources.

Huheey, J. E. (1983) *Inorganic chemistry*. Harper, Cambridge, Md.

Portefield, W. W. (1984) *Inorganic chemistry*. Addison-Wesley, Reading, Mass.
Purcell, K. F. & Kotz, J. C. (1977) *Inorganic chemistry*. Saunders, Philadelphia.
Chapter 4 describes the application of TASOs to metal complexes particularly
   clearly.

Text references
Jørgensen, C. K. (1962) *Absorption spectra and chemical bonding in complexes*.
   Pergamon, New York. (Lattice energy in Fig. 8.6).
Stout, W. J. and Reed, S. A., *J. Amer. Chem. Soc*, **76**, 5279 (1954); (M-F distances).

9

Complexities of covalent bonding

9.1 INTRODUCTION

Chapter 5 examined simple covalent bonding, involving the overlap of s and p orbitals. Interactions with d orbitals produce the molecular orbitals involved in transition-metal donor-acceptor complexes (Chapter 8). Similarly, d orbitals on main group elements are involved in many of the covalent compounds discussed in the present chapter. Other complicated bonding patterns are the result of the interaction of orbitals on many different atoms.

9.2 d ORBITALS OR NOT?

The chemical contrasts between the non-metals of the second period (Si–Cl) and those of the first period (C–F) are very marked. Differences in valency and coordination number are particularly prominent. Thus, we find SF_6 and PF_5 as the highest-valent fluorides of S and P, compared to OF_2 and NF_3. Silicon forms SiF_6^{2-}, a complex with no analogue among stable carbon compounds. The existence of these compounds is normally explained from the electronic structures of the second-row elements. Sulphur, for example, contains two electrons in the 3s and four in the 3p orbitals of its valence shell. As with oxygen, these can produce sp^3 hybrid orbitals, forming two σ-bonds and two lone pair orbitals (as in Me_2S). Sulphur also has five 3d orbitals with the same principal quantum number as its 3s and 3p orbitals. In isolated atoms, of course, these orbitals are empty, and as the nuclear charge increases along the periodic table they remain so until the transition metals are reached. However, their involvement in bonding is used to explain the higher valency and coordination number of S compared to O. We have already assumed this for the structures of such molecules as SF_4 and PF_5 when applying the VSEPR approach (section 5.6). Sets of hybrid orbitals accommodate the pairs of electrons, and d orbitals are needed if more than four pairs are used, in sp^3d, sp^3d^2, or sp^3d^3 hybrids. Interaction with the d orbitals of phosphorus is also used to explain the behaviour of metal complexes containing PR_3 ligands (section 8.3.2).

These assumptions gloss over one serious problem: how can the d orbitals be involved if they are too high in energy to contain electrons? This is not too serious in $M(PR_3)_n$ complexes, where the d orbital involvement can be a modest addition to a dative σ-bond formed from a sp^3 lone pair orbital. However, in the VSEPR description of SF_6 and similar compounds, the d orbitals are involved equally with the s and p orbitals.

9.2.1   Bonding in XeF_2

This problem came to a head when the xenon fluorides were discovered. It is easy to explain the linear geometry of XeF_2 by the VSEPR method. It contains five electron pairs and so is predicted to have trigonal bipyramidal geometry with three equatorial lone pairs and two axial F–Xe–F bonding pairs in sp^3d hybrid orbitals. However, the surprise in the chemical community at these compounds of the 'inert' gases led to the suggestion that their bonding might reflect their supposed instability.

A model was therefore proposed which did not require d orbital involvement. Consider the interaction between two F-p_z orbitals (one electron each) and a full Xe-p_z orbital (Fig. 9.1a, taking the molecular axis as z, for convenience). These three orbitals give three molecular orbitals: bonding, non-bonding and anti-bonding (Fig. 9.1b). With four electrons in the system, the first two are filled, and the presence of one bonding pair gives a 0.5 bond order for each Xe-F bond. On this view, XeF_2 is a three-centre/four-electron system (3c-4e), exactly as suggested for hydrogen bonding and secondary bonding (sections 7.4–7.5).

However, this should not be the end of the story. We can now allow the Xe empty d orbitals to interact with these molecular orbitals. Consider specifically the d_{z^2} orbital (Fig. 9.2a). Its symmetry† means that it will interact only with ψ_{nb}. This interaction produces two new orbitals, ψ_1 and ψ_2, respectively more and less stable than ψ_{nb}. Just how much ψ_1 gains in stability depends on the strength of the interaction with the d_{z^2} orbital. If this interaction is modest, the pair of electrons in the ψ_{nb} orbital will become slightly more stable. This will increase the Xe–F bond order slightly (Fig. 9.2b). However, if the Xe-d_{z^2} orbital interacts strongly with ψ_{nb}, the resulting bonding orbital (ψ_1) may be almost as stable as ψ_b (Fig. 9.2c). The molecule now has two bonding pairs, giving a bond order of 1.0 for each Xe–F bond. This is equivalent to forming the bonds in XeF_2 by promoting an electron to the Xe-d_{z^2} orbital and constructing sp^3d hybrids by the VSEPR approach.‡

This treatment shows that the two alternative approaches to XeF_2 are not mutually exclusive. The 3c-4e model and the VSEPR model are at each end of a continuous range, and the position of XeF_2 along this continuum depends on the strength of the d orbital interaction. Its position can be discovered only from experimental evidence. Of course, different molecules will differ in the extent of their d orbital involvement and therefore in their bond order.

† g-symmetry in the terminology of group theory.
‡ Looking at the trigonal bipyramidal sp^3d orbitals in detail, we find that the s orbital and two of the p orbitals give rise to the equatorial orbitals, while the remaining p orbital and the d orbital produce the axial orbitals.

(a)

$$\psi_1 \qquad\qquad \psi_2 \qquad\qquad \psi_3$$

(b)

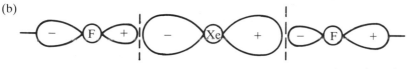

$$\psi_{ab} = \psi_1 - \psi_2 + \psi_3$$

$$\psi_{nb} = \psi_1 - \psi_3$$

$$\psi_b = \psi_1 - \psi_2 + \psi_3$$

(c)

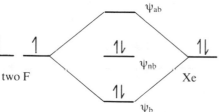

Fig. 9.1 — Bonding in XeF_2 through p-orbitals only (three-centre/four-electron model) (a) atomic orbitals, (b) molecular orbitals, (c) energy diagram.

9.3 EXTENT OF d ORBITAL INTERACTION

We next examine a variety of compounds to assess the amount of d orbital involvement in each. We have one guide in this, a study published as long ago as 1956, which has dropped rather out of sight since then. Craig & Magnusson (1956) considered the effect on d orbital energy of the approach of electron-attracting groups. These groups were modelled very simply as positive charges, which are ideal electron attractors. They were placed in an octahedral arrangement around P or S (Fig. 9.3a). The effect on the d orbitals is dramatic. An isolated 3d orbital on a

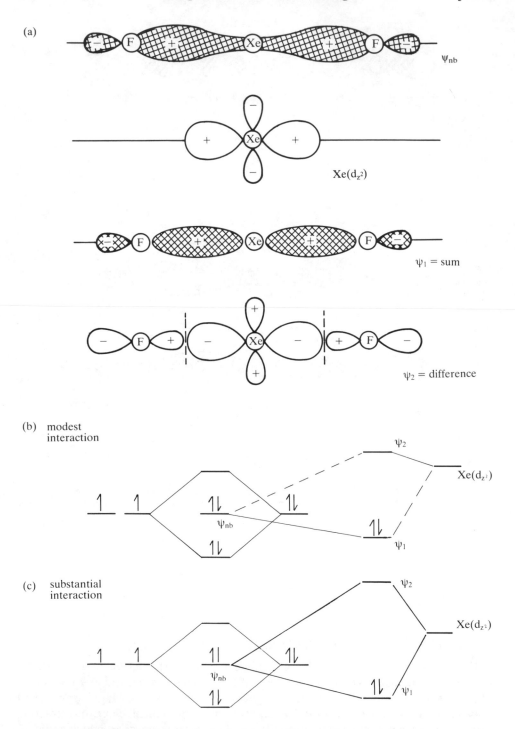

Fig. 9.2 — Interaction of ψ_{nb} orbital of XeF_2 (3c-4e model) with Xe-d_{z^2} orbital (a) geometry, (b) energy diagram with modest interaction, (c) energy diagram with substantial interaction.

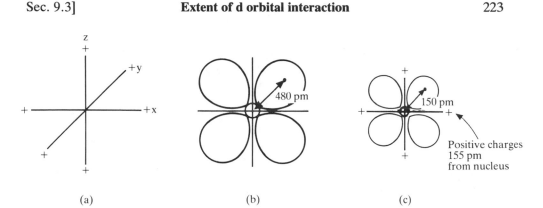

Fig. 9.3 — Effect of positive charges on a d-orbital (a) arrangement of charges, (b) electron density contour for an isolated d-orbital, (c) electron density contour for a d-orbital with six adjacent positive charges.

phosphorus atom is very diffuse, with maximum electron density about 480 pm from the nucleus (Fig. 9.3b). Clearly, its overlap with the bonding orbital of an adjacent atom will be very small indeed. With six $+1$ charges placed 155 pm from the phosphorus (the length of a P$-$F bond), the 3d orbital shrinks so that its maximum is only 150 pm from phosphorus (Fig. 9.3c). Overlap with a fluorine p orbital is now excellent. Naturally, this contraction of the d orbital is accompanied by an increase of the stability of an electron *in the molecule*, because it is being attracted both by the phosphorus nuclear charge and by the neighbouring positive charges. The charges used in this model are more electron-attracting even than fluorine atoms, and so the precise values of orbital size and energy must not be taken to represent the exact situation in PF_6^-. These calculations show clearly that the diffuse d orbitals are very sensitive to the presence of electron-attracting atoms.

The implications are clear. For the d orbitals on a second row (or later) non-metal to be used extensively, this atom must be surrounded by electronegative groups. This agrees very well with the observed chemical behaviour. The involvement of d orbitals can be judged experimentally in three ways by (i) the geometry of molecules apparently using sp^3d^n hybridization, especially when they contain lone pairs, (ii) the lengths of bonds, (iii) the ability to form compounds in high oxidation states.

9.3.1 Molecular geometry

The linear XeF_2 molecule and the square planar XeF_4 satisfy both the 3c-4e and the VSEPR models. However XeF_6 gives a chance to distinguish them (Fig. 9.4). It could be a regular octahedron, with each p orbital involved in a 3c-4e bond. Alternatively, using the VSEPR approach, its seven electron pairs will be in sp^3d^3 hybrid orbitals. By analogy with IF_7, its geometry should then be pentagonal bipyramidal. Experimentally, neither is found, though the observed structure shows that the VSEPR model gives the better description. The prediction of a pentagonal bipyramid overlooks an important difference between a bound atom and a lone pair. The latter can move much more rapidly from place to place. As a consequence in gaseous XeF_6, although the lone pair causes the molecule to distort, the form of this

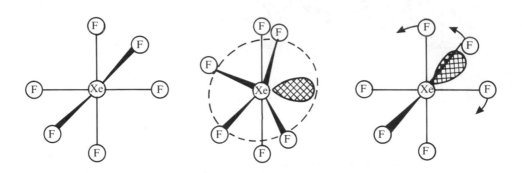

Fig. 9.4 — Possible geometries for XeF_6 (a) regular octahedron: 3c-4e model, (b) pentagonal bipyramid with one vacant position: VSEPR model, analogous to IF_7, (c) observed geometry: distorted octahedron with lone pair moving rapidly from one face to another.

distortion changes rapidly. Instantaneously, the lone pair projects through one face of an XeF_6 octahedron, enlarging the three adjacent F−Xe−F angles. However, the lone pair moves rapidly from one face to another of the octahedron as the molecule vibrates, affecting different groups of three F atoms.†

The geometries of some other non-metal compounds can be explained by a 3c-4e model, such as the regular octahedron of SF_6 and the trigonal bipyramid of PF_5 (with equatorial sp^2 bonds and an axial F−P−F 3c-4e system). However, the VSEPR approach is equally satisfactory for them, and it is much more useful for molecules containing lone pairs. IF_5 resembles SF_6 except for its lone pair. Although a 3c-4e F−S−F system is apparently reasonable in the first, the corresponding (lone pair)−I−F system is chemically less sensible. SF_4 and PF_5 are also very similar, apart from the former's equatorial lone pair; the geometry of SF_4 is easy to understand only if all the orbitals are roughly equivalent in character.

Such examples are so numerous, that it is hard to find any exceptions. However, one small group of compounds should be mentioned. TeX_4 and SbX_3 compounds with X = Cl, Br, I will add X^- to give TeX_6^{2-} and SbX_6^{3-}, though when X = F they form only TeF_5^- and SbF_5^{2-}. The two pentafluoride ions have the square-pyramidal geometries expected for six electron pairs (including one lone pair); they are the same shape as IF_5 and XeF_5^+. In contrast, the MX_6^{n-} ions are all regular octahedra. Using the VSEPR description, six of their seven electron pairs are assigned to M−X bonds, and the remaining one is considered to be in a pure s orbital. This spherically symmetrical s orbital is 'stereochemically inactive', so the six bonding pairs of electrons take up regular octahedral geometry.

Bearing in mind that these MX_6^{n-} ions form with only the less electron-attracting

† The detailed analysis of gas-phase XeF_6 shows that its geometry is very complex. The original report should be consulted for full details: Bartell & Garvin, (1968).

halogens, they can alternatively be described as compounds in which the d orbital participation is very small. On this view, they contain three sets of 3c-4e bonds using each p orbital of the central atom. As will be seen in the next section, their bond lengths, as well as their geometries, are consistent with this interpretation. They are therefore the reverse of XeF_6, for which the geometry supports d orbital involvement.

9.3.2 Bond length evidence

Ideally, bond lengths can easily distinguish between the bonds of order 0.5 or 1.0 in the two descriptions of XeF_2. This should be possible by using either the precise relationship (9.1)

$$\log_{10}(n) = (r_o - r)/c \tag{9.1}$$

or the simple approximation that halving the bond order lengthens the bond by about 30 pm (section 5.8). In reality, it is very difficult to establish with total confidence the length of the standard single bond to which others can be related. However, it is often easier to compare pairs of similar compounds and use the difference in their bond lengths to understand their bonding. For the xenon fluorides, it is notable that the Xe−F bonds become shorter as the number of fluorine atoms increases: XeF_2 200 pm, XeF_4 193 pm, XeF_6(gas) 189 pm. The length of a single Xe−F bond is not easy to estimate, but the distance in Xe-F$^+$ may give a reasonable value as this ion is isoelectronic with IF. In F-Xe$^+$..$[Sb_2F_{11}]^-$, the Xe-F distance is 184 pm and it is 190 pm in $[Xe_2F_3]^+$ $[AsF_6]^-$ (in which the cation can be described as $[XeF]^+$ ...F$^-$...$[XeF]^+$); the unknown gas-phase distance in $[Xe-F]^+$ is probably slightly shorter than these. These distances suggest that the bond order in XeF_6 is close to 1.0, with extensive d orbital involvement. XeF_2 appears to have a rather lower bond order and therefore less d orbital contribution.

By contrast, the I−Cl distances in $PhICl_2$ (245 pm), ICl_2^- (255 pm), and ICl_4^- (252 pm, mean) are substantially longer than in ICl (230 pm). The difference between ICl and ICl_2^- is large enough to suggest that in ICl_2^- and ICl_4^- the bond order is close to 0.5, with 3c-4e bonds. The effect of increasing the number of I−Cl bonds is clearly the opposite of increasing the number of Xe−F bonds.

A second comparison between a compound that may involve d orbitals, and one that certainly does not, was glimpsed in section 7.3.3.3, when $Me_3N\rightarrow O$ was discussed. It became clear that the dative N→O bond corresponds approximately to a single bond. The corresponding phosphine oxides, such as Me_3PO, can be described in the same way without involving d orbitals; the lone pair on P donates to the empty orbital of the O-atom. This leaves the phosphorus tervalent. Alternatively, if an electron is promoted to a d orbital, the phosphorus becomes pentavalent and forms a double bond to the oxygen. The bond distances in Me_3PO show very clearly that the second picture is better. The C−P distance (181 pm) gives a value of 104 pm for the phosphorus single bond radius, leading to a prediction of 170 pm for a P-O single bond (using the values in Appendix A.1). The observed P−O distance is 148 pm, and the shrinkage of 22 pm is in the range anticipated for the change from a single to a double bond.† This strong P−O bond underlies the difference in

---

† The radius difference between O(single) and O(double) is 11 pm (Appendix A.1). We do not have an independent value for this difference with P, but can expect a shrinkage of 10–15 pm (section 5.7.1).

chemistry between amine and phosphine oxides. The former are thermally unstable and are strong oxidizing agents, while the latter are stable and non-oxidizing. Indeed, the formation of molecules of general formula $R_3P = O$ is one of the dominant themes of phosphorus chemistry. One other such compound, F_3PO, is considered in section 9.4.

The final application of bond length evidence is to the TeX_6^{2-} and SbX_6^{3-} ions with stereochemically inactive lone pairs, and their contrast with TeF_5^-, and SbF_5^{2-}. The latter are square pyramidal, and the M$-$F distances (Te$-$F 194 pm; Sb$-$F 203 pm) correspond approximately to those expected for single bonds, e.g. Sb + F 199 pm (Appendix A.1). The predicted single bond lengths and the observed distances for the chlorides and bromides are as follows:

Ion	Predicted (pm)	Observed (pm)	Difference
$TeCl_6^{2-}$	234	254	20
$SbCl_6^{3-}$	243	264	21
$SbBr_6^{3-}$	261	280	19

Clearly, these ions have bond orders closer to 0.5 than 1.0. Thus, their dimensions confirm the deductions made from their geometries, that they have 3-centre bonding with little d orbital contribution.

9.3.3 High valency compounds

Involvement of d orbitals on non-metals also influences their valency. The formation of high-valent compounds is clearly restricted to interactions with electronegative elements. Thus in the compounds of sulphur in their highest oxidation states, SH_2, $S(CH_3)_2$, SBr_2, SCl_4, SF_6, and SO_3, only with F and O is the hexavalent state reached. Exactly the same pattern is found, for example with iodine halides, the highest-valent compounds being IF_7, I_2Cl_6, and IBr. The chemistry of xenon is particularly interesting because of the *virtual* non-existence of compounds other than fluorides and unstable oxides, i.e. with H, C, S, etc. as attached groups the valency does not exceed zero. Recently some notable exceptions to this pattern have been discovered as stable ions : $[F\{Xe(NSO_2F)_2\}_2]^+$, $[(C_6F_5)Xe]^+$, $[XeO_3Cl_2]^{2-}$. In these compounds, even though the atoms directly bonded to Xe are not very electronegative, their *effective* electron-attracting power has been greatly enhanced by the attached groups. Thus these substituents behave as if they were fluorine atoms.

This pattern is understandable in the light of the effect of electronegative groups on the d orbitals. The orbitals are stabilized for bonding, so that electrons can be promoted into them, increasing the valency of the central non-metal. This chemical behaviour, like the evidence of geometry and of bond lengths, confirms that d orbitals *are* extensively involved in compounds of non-metals in higher oxidation states with electronegative elements or groups. This includes the xenon fluorides for which the 3c-4e model was originally proposed. However, d orbital involvement is not an all-or-nothing matter. In compounds such as the iodine chloride ions, the bonding conforms closely to the 3c-2e model.

9.4   DATIVE BONDING TO d ORBITALS

Much of the distinctive chemistry of second row elements is attributed to their d orbitals. The preceding section examined their involvement in higher valency compounds. An increase in valency is not possible for group 14 elements such as silicon, simply because all their electrons are already involved in bonding. However, the silicon d orbitals can be used in dative bonds, for example giving $[SiF_6]^{2-}$ from SiF_4. $[SiF_6]^{2-}$ is isoelectronic with $[PF_6]^-$ and SF_6, and all three can be described by sp^3d^2 hybridization. Many similar compounds are known, especially of the group 14 elements Si, Ge, Sn, following the patterns of donor-acceptor bonding described in Chapter 7. However, in some compounds of Si the d orbitals appear to be involved in π- rather than σ-bonding.

9.4.1   Dative π-bonding to d orbitals

A number of silicon compounds have overall structures that are nearly the same as the corresponding carbon compounds, but which differ in detail (Ebsworth, 1987). A good example is trisilylamine $(SiH_3)_3N$. This is planar at nitrogen, in contrast to $(CH_3)_3N$, which has the pyramidal shape expected for an sp^3 hybridized atom. In addition, photoelectron spectroscopy shows that the nitrogen lone pair is more tightly bound than in $(CH_3)_3N$.

This behaviour is best explained by donation from this lone pair to the empty d orbitals on silicon. In order to achieve satisfactory overlap, the nitrogen lone pair has to be in a p orbital, rather than a sp^3 hybrid, i.e. the geometry at sp^2 nitrogen must be trigonal planar. Donation $(p_\pi \rightarrow d_\pi)$ into the empty d orbitals of silicon then takes place (Fig. 9.5). The effect on the Si−N bond distance is significant, though not very

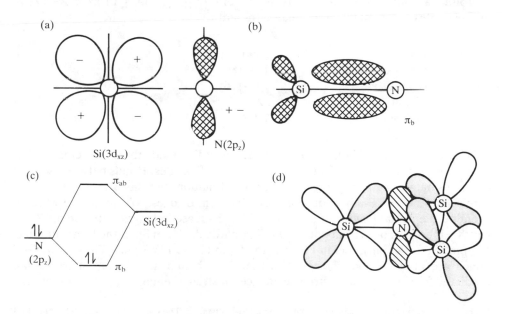

Fig. 9.5 — $p_\pi \rightarrow d_\pi$ donation in $(SiH_3)_3N$ (a) atomic orbitals for one bond, (b) bonding interaction, (π_b), (c) energy levels, (d) view of all three bonds.

large. From the carbon-based covalent radii of N and Si, an Si−N distance of 179.2 pm would be predicted, compared to the observed value of 173.4 pm.† Exactly the same behaviour is seen with Si−O bonds. The Si−O−Si bond angle in $(SiH_3)_2O$ is 142°(solid) or 144°(gas). This implies that the hybridization is intermediate between sp (linear) and sp^2 (120° Si−O−Si angle). As a result the oxygen lone pairs can interact with the Si d orbitals. This π interaction involving the d orbitals is similar in character to that already described for metal complexes (section 8.3).

Other SiH_3 or Me_3Si compounds, with Si bound to O or N, show the same effects. It is also clear that the Si−O bonds in silicates are shorter and stronger than would be expected for single bonds (e.g. 160.7 pm in quartz, compared to a prediction of 175 pm). In contrast to the strength of these nominal single bonds, simple π-bonds to second row elements such as silicon are generally weak; thus, the silicon analogues of ethene are stable only with bulky groups attached (Greenwood & Earnshaw 1984, p.419). This is important in relation to the general chemical behaviour of silicon, well illustrated by the contrast between CO_2 and SiO_2. The first is a monomer containing both σ- and π-bonds, the second an infinite polymer with nominal Si−O single bonds in reality strengthened by $p_\pi \rightarrow d_\pi$ bonds. For

$$Si(-O-..)_4 \quad \text{to become} \quad O=Si=O,$$

requires two strong Si−O single bonds to be replaced by weak Si−O π-bonds, which is energetically very unfavourable.

This $p_\pi \rightarrow d_\pi$ bonding is particularly important for silicon, simply because its d orbitals can only *accept* electrons. Good evidence also exists for such bonding with other non-metals, though their d orbitals can also form covalent σ-bonds. Naturally, this π-bonding is most prominent in compounds with electronegative atoms when the d orbitals are most accessible. Thus the dimeric $[X_2O_7]^{n-}$ ions all have X−O−X bridge angles substantially larger than tetrahedral:

Ion	Angle(°)
$[O_3Si\text{-}O\text{-}SiO_3]^{6-}$	133–180
$[O_3P\text{-}O\text{-}PO_3]^{4-}$	134–156
$[O_3S\text{-}O\text{-}SO_3]^{2-}$	124
$O_3Cl\text{-}O\text{-}ClO_3$	118.6

In the Si−N and Si−O compounds already discussed, the involvement of d orbitals is revealed directly by the unusual Si−X−Si angles. Bonds between second row non-metals and F may also contain p_π-d_π donation, but this is more difficult to detect because the F-atoms do not normally form bridges. We therefore have to examine the actual lengths of the bonds. The decrease in length between an X−C and X−F bond should be about 15 pm (Appendix A. 1). In SiF_4, the bond length is 154 pm (Si-C 186 pm), and in PF_3 it is 157.0 pm (P−C 185.5 pm). Thus both X−F bonds are about 30 pm shorter than the X−C bonds, i.e. 15 pm shorter than predicted for single bonds − strong evidence for strengthening by p_π-d_π interactions.

---

† These figures may slightly exaggerate the expected Si-N distance. The C-N distance in $(CH_3)_3N$ (145.4 pm) is shorter than that given by the covalent radii in Appendix A.1 (146.9 pm), perhaps because of hybridization changes.

With an additional electronegative group in $F_3P = O$, the P—F distance is shorter still (152.4 pm), while the $P = O$ distance (143.6 pm) is significantly shorter than that in $Me_3P = O$ (148 pm).

9.5 BONDING BETWEEN HIGHLY ELECTRONEGATIVE ELEMENTS

The preceding section implied that covalent bonds involving very electronegative atoms are unusually strong. This is correct, as long as the second atom in the bond is not itself highly electronegative. The best illustration of the opposite effect comes from compounds of oxygen and fluorine, in particular F_2O_2, which can be compared to H_2O_2 (Fig. 9.6).†

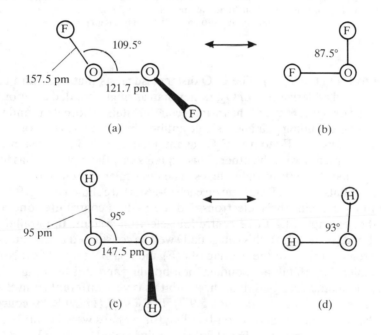

Fig. 9.6 — F_2O_2 and H_2O_2: views parallel (a, c) and perpendicular (b, d) to the O—O bond.

Both F_2O_2 and H_2O_2 are non-planar, with dihedral angles of about 90° between the two $X-O_2$ halves of the molecules (Figs 9.6b,d). This is their first surprising feature. Dihedral angles of $\pm 60°$ or 180° are expected for the bond between two sp^3 atoms, depending on the bulk of the substituents. Fig. 9.7 shows how these angles lead to the maximum separation of the electron pairs in a $-(CH_2)_n-$ chain. Clearly, other interactions are important in F_2O_2 and H_2O_2. The bond lengths in F_2O_2 are also unexpected. The O—O distance is extremely short compared to H_2O_2, and indeed is hardly longer than in O_2 (120.7 pm). By contrast, F—O is very long

† The approach to the unusual bonding in F_2O_2 was originally proposed by Spratley & Pimentel (1966).

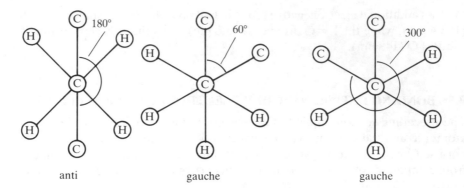

Fig. 9.7 — Normal conformations about the C—C bond in a −(CH$_2$)n− chain, viewed down the C—C bond. The *anti* conformation (dihedral angle 180°) is the most favoured, but *gauche* conformations are also energetically satisfactory.

compared to F$_2$O (140.9 pm). The O-O distance suggests that F$_2$O$_2$ and H$_2$O$_2$ can best be described as derivatives of O$_2$, rather than as single-bonded compounds with sp^3 hybridized oxygen atoms. O$_2$ has π-molecular orbitals in both its x- and y-planes, each containing a bonding pair and a single antibonding electron (section 5.9.1). We can therefore view one F-atom in F$_2$O$_2$ as interacting with one π-system, and the second F as interacting with the other. This explains why the molecule has the two F-O-O planes almost exactly at right angles. The molecular orbitals constructed from the three p orbitals (one on F and one on each O-atom) are shown in Fig. 9.8. As each π-system in O$_2$ contains three electrons and the F-atom contributes one more, we have another example of a three-centre/four-electron system. In parallel with the hydrogen bond (section 7.4), this can equally well be considered as the interaction of the three atomic orbitals or the overlap of p(F) and π_{ab}(O$_2$) (Fig. 9.8e). Nominally, the three molecular orbitals are bonding, non-bonding and anti-bonding (Fig. 9.8b). However, the spatial distribution of these orbitals is very different from that in, for example, the π orbitals of CO$_2$ (section 5.9.2). In F$_2$O$_2$ ψ_b (Fig. 9.8c) is concentrated between the two O-atoms and is non-bonding (or perhaps weakly anti-bonding) in the F-O bond. In contrast, ψ_{nb} (Fig. 9.8d) is mainly located between F and O, where it is weakly bonding; it has some O−O anti-bonding character.

The description as a p(F)-π_{ab}(O$_2$) interaction (Fig. 9.8e) is rather cruder. The two electrons in these orbitals occupy the lower molecular orbital which is essentially the same as ψ_{nb} of the three-centre description: bonding in F−O and weakly anti-bonding in O−O. This description assumes that the O−O π_b orbital is unaffected.

The three-centre approach to H$_2$O$_2$ (Fig. 9.9) involves the same orbitals. However, the different electron-attracting power of F and H leads to the key difference between F$_2$O$_2$ and H$_2$O$_2$. In H$_2$O$_2$, the ψ_b orbital has high electron density between H and O, because the 1s(H) orbital is not as closely bound to the H-nucleus as the 2p(F) orbital is to the F-nucleus. The ψ_{nb} orbital is now strongly anti-bonding between the oxygen atoms, leading to a long O−O bond. Finally, it is worth remembering that such variation between relatively symmetrical and very unsymme-

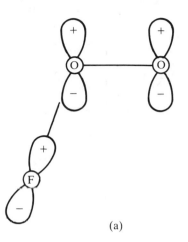

(a)

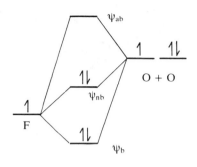

(b)

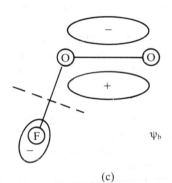

ψ_b

(c)

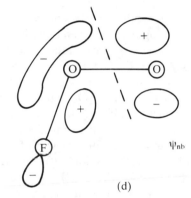

ψ_{nb}

(d)

(e)

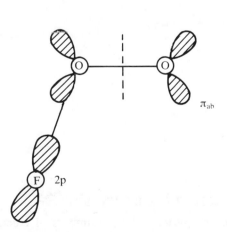

π_{ab}

F   2p

Fig. 9.8 — Molecular orbital system in one F−O−O plane in F_2O_2. (a) atomic orbitals, (b) energy levels, (c) ψ_b, (d) ψ_{nb}, (e) Alternative view: two orbital interaction $p(F) + \pi_{ab}(O_2)$.

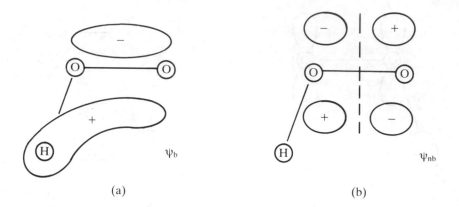

Fig. 9.9 — Molecular orbitals in one H$-$O$-$O plane of H$_2$O$_2$ (a) ψ_b, (b) ψ_{nb}.

trical bonds, depending on the nature of the atoms involved, is typical of other 3c-4e systems including the hydrogen bond itself.

This approach can be extended to a whole series of weakly bonded F$-$O$_2$ complexes, such as F$_2$O$_4$, representable as (I); it also explains why O$_2$ and NO do not dimerize or polymerize, despite their unpaired anti-bonding electrons. It remains unclear why in H$_2$O$_2$ and F$_2$O$_2$, the 2s(O) orbitals do not become involved in bonding to F or H, which would lead to standard sp^3 hybrid orbitals (or their equivalent in σ-molecular orbitals). Only a tentative solution can be offered for this problem: that the strong electron-attracting power of the *two* O-nuclei (due to their high nuclear charge and low electron shielding) leads to the 2s orbitals being contracted, so that their overlap with 2p(F) or 1s(H) is small. The anomalous behaviour of F$_2$O$_2$ and H$_2$O$_2$ can then be seen as the direct consequence of the juxtaposition of the two electronegative atoms, paralleling the modification by the six F-atoms in SF$_6$ of the d orbital energies of the sulphur atom.

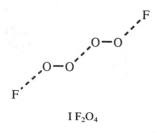

I F$_2$O$_4$

9.6   ELECTRON-DEFICIENT COMPOUNDS

From compounds with exceptionally large numbers of electrons, we turn to the opposite situation: compounds with too few electrons for standard two-electron bonds between each pair of bonded atoms. These compounds are conventionally called 'electron deficient', although this is rather a misnomer. In reality, these compounds take up geometries in which the number of bonding molecular orbitals matches the available electrons. Recent developments in the description of electron-deficient compounds enable the structures and formulae of some remarkable molecules to be correlated and understood.

Boron is the element above all that forms electron-deficient compounds. It possesses only three outer electrons but has four low-lying valence orbitals ($2s$, $2p_x$, $2p_y$, $2p_z$), and is sufficiently electronegative to form only covalent bonds. Other group 13 elements can lose electrons to give ionic species (Al^{3+} etc.), but the ionization energies for boron are too high for this to occur. Thus the chemistry of boron is largely an exploration of how it fills its fourth orbital. We have already encountered two methods: $p_\pi \rightarrow p_\pi$ donation as in BF_3 (section 5.9.4) and donor-acceptor bonding as in $Me_3N \rightarrow BMe_3$ (section 7.3).

9.6.1 Diborane

The simplest stable boron hydride, diborane (B_2H_6), shows a third method. Although an equilibrium with the monomer BH_3 is set up if gaseous B_2H_6 is heated, the proportion of monomer is very low, and it rapidly reacts with the dimer to give other hydrides. The structure of diborane (Fig. 9.10) shows that the BH_3 units

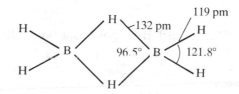

Fig. 9.10 — Structure of diborane.

dimerize through bridging hydrogen atoms. In the absence of any other electrons, the vacant orbital on each boron is filled by donation from the *bonding* pair of electrons in a B−H bond (II).

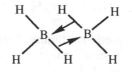

II B_2H_6 – donation from bond pairs

Because B_2H_6 is symmetrical, its bonding is more clearly described by starting from the atomic orbitals rather than from BH_3 units (Fig. 9.11). Each boron can be taken as sp³ hybridized,† forming two standard electron-pair bonds to the terminal H-atoms. These use eight electrons from the total of twelve in the whole molecule. The bridges contain two equivalent sets of three orbitals: an sp³ hybrid on each boron and the 1s orbital of the bridging hydrogen (Fig. 9.11a). These each form three molecular orbitals, ψ_b, ψ_{nb} and ψ_{ab}, and the four electrons fill the two ψ_b orbitals (Fig. 9.11b–c). Both bridges can be described as three-centre/two-electron systems,

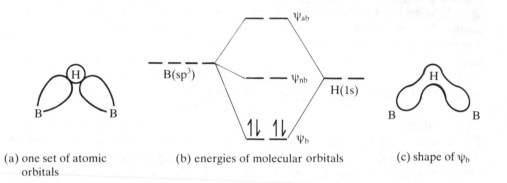

(a) one set of atomic orbitals   (b) energies of molecular orbitals   (c) shape of ψ_b

Fig. 9.11 — Molecular orbitals of B–H–B bridges in B_2H_6 (a) one set of atomic orbitals, (b) energy diagram (both bridges), (c) shape of ψ_b.

with bond strengths of 0.5 for each B–H bond (one bond pair for two bonds in each bridge).

9.6.2 Higher boranes
The structure of B_2H_6 is remarkable compared to hydrocarbons such as C_2H_6, but the higher boranes can only be called bizarre! The following formulae of the first members show no apparent logical progression and their structures are just as baffling:

B_2H_6, [no stable B_3 species], B_4H_{10}, B_5H_9, B_5H_{11}, B_6H_{10}, B_6H_{12}.

Furthermore, to add to this complexity, families of related species exist:

Anions: $[B_6H_6]^{2-}$ and other $[B_nH_n]^{2-}$ ions (n = 7–12), $[B_3H_8]^-$, $[B_5H_8]^-$, $[B_5H_{10}]^-$, $[B_{10}H_{12}]^{2-}$, $[B_{10}H_{14}]^{2-}$

Boranes and borane anions containing hetero-atoms, especially carbon (carboranes and carborane anions): CB_5H_9, $C_2B_3H_5$, $C_2B_{10}H_{12}$, $[C_2B_9H_{11}]^-$, SB_9H_9, $[NB_9H_{13}]^-$

Boranes combined with donors: $B_3H_7(NMe_3)$, $B_5H_9(PMe_3)_2$, $B_{10}H_{10}(NCMe)_2$

† The four hybrid orbitals need not be equivalent, and the large external H–B–H angle, compared to the normal tetrahedral value, suggests that these orbitals may be closer to sp² in character, with the bridging orbitals having a larger p-component.

The first step in understanding this confusion is the recognition that isoelectronic compounds have similar structures. Thus, BH^- or BH_2 can be replaced by CH, relating $[B_6H_6]^{2-}$ to $[CB_5H_6]^-$ and $C_2B_4H_6$; B_6H_{10} to CB_5H_9 and $C_2B_4H_8$. Extending this, N can also replace BH^- and PH or S can replace BH_2^-, so that $[NB_9H_{13}]^-$ and $[B_{10}H_{14}]^{2-}$, SB_9H_9 and $[B_{10}H_{10}]^-$, are equivalent pairs (though not many species with N, P, or S hetero-atoms are known, probably because of a lack of systematic study). Similarly, a donor L corresponds to two electrons, giving $L_2B_{10}H_{12}$ and $[B_{10}H_{14}]^{2-}$ (or the unknown $[B_{10}H_{12}]^{4-}$) as corresponding species.

9.6.2.1 Borane families

With these links, it becomes possible to correlate the formulae and structures of the boranes and their derivatives. It was shown by Wade (1976) that they can be systematically grouped into families. Each family starts with a borane anion with the structure of a regular polyhedron, formula $[B_nH_n]^{2-}$. Polyhedra with $n = 6$–12 are known, and the related carborane $C_2B_3H_5$ exists for $n = 5$ (Fig. 9.12). These are the *closo*-borane anions. Their most important feature is the distribution of electrons in the polyhedron. Taking $[B_6H_6]^{2-}$ as an example, this ion has 26 electrons. Each B−H bond contains one pair, leaving 14 electrons (7 pairs) for the B_6 octahedron. These are known as the *skeletal* electrons.

As will be seen below, precisely 14 electrons are needed to fill the bonding orbitals of the B_6 cluster. More generally, an *n*-atom closed polyhedron requires $(n + 1)$ pairs of electrons for stability. The crucial step in understanding the borane structures is to relate these closo-borane anions to the neutral boranes, which have open rather than closed polyhedra of boron atoms.

> Removal of one vertex (as a formal step, not a chemical reaction) from a *closo*-borane gives a *nido*-borane (Latin *nidus*, a nest). *This is stable with the same number of skeletal electrons as the parent polyhedron.*

Thus, from $[B_6H_6]^{2-}$ we remove one vertex with only the pair of electrons in the B−H bond, i.e. BH^{2+}, obtaining $[B_5H_5]^4$ (an octahedron with one missing vertex). This picks up four protons around its open face to give the stable *nido*-B_5H_9 (Fig. 9.13e). The protons form B−H−B bridges across each open B−B edge. It is easy to confirm the number of skeletal electrons in B_5H_9. The molecule contains 24 electrons in all; five pairs occupy the *terminal* B−H bonds, leaving 14 skeletal electrons, as for $[B_6H_6]^{2-}$.

The process can be repeated to give an *arachno*-borane from a *nido*-borane (*arachne*, Greek for spider's web, indicating a more open structure). $[B_5H_5]^{4-}$ then gives $[B_4H_4]^{6-}$, which adds protons to become B_4H_{10} (Fig. 9.13f). The second missing vertex is adjacent to the first, so that the boron atoms in B_4H_{10} are arranged in two linked triangles. The six extra hydrogen atoms are arranged around the open face. Four of them form B−H−B bridges and the others are in B−H bonds on the end boron atoms. An important distinction is made between these and the original B−H bonds. Every boron atom carries one H-atom pointing out from the polyhedron. These are known as the *exo* H-atoms. Other terminal H-atoms, such as those on the end boron atoms in B_4H_{10}, are *endo* H-atoms. In counting the skeletal electrons,

(a)

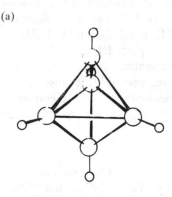

(b)

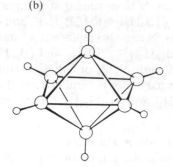

(c)

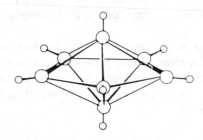

(d)

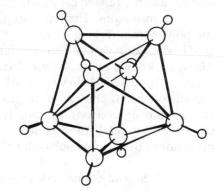

(e)

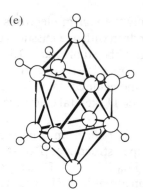

(f)

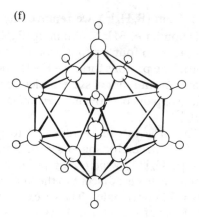

Fig. 9.12 — *Closo*-borane anions (a) $[B_5H_5]^{2-}$, (b) $[B_6H_6]^{2-}$, (c) $[B_7H_7]^{2-}$ (inferred), (d) $[B_8H_8]^{2-}$, (e) $[B_{10}H_{10}]^{2-}$, (f) $[B_{12}H_{12}]^{2-}$. B_9 and B_{11} *closo*-boranes are also known.

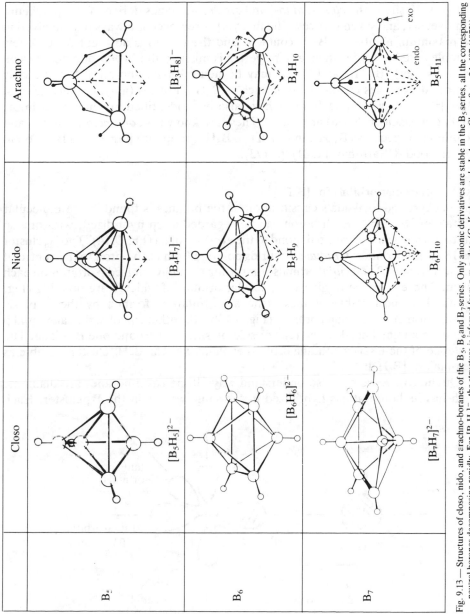

Fig. 9.13 — Structures of closo, nido, and arachno-boranes of the B$_5$, B$_6$ and B$_7$ series. Only anionic derivatives are stable in the B$_5$ series, all the corresponding neutral boranes decomposing rapidly. For [B$_4$H$_7$]$^-$, the structure is inferred from n.m.r. data (G. Kodema et al., *J. Amer. Chem. Soc.*, **94**, 407 (1972)).

the electron pairs of the *endo* H-atoms are included, but not those of the *exo*-bonds. Thus in B_4H_{10}, of the 22 total electrons, 8 are in *exo* $B-H$ bonds, again leaving 14 skeletal electrons.†

Fig. 9.13 shows the *closo*-, *nido*-, and *arachno*-boranes derived from polyhedra with five, six, and seven vertices. The first vertex removed is always one involved in the maximum number of $B-B$ contacts, and the second is adjacent to it. A few examples are known in which a third vertex is missing, the *hypho*-boranes (Greek *hyphe* = net). $B_{10}H_{18}$ and B_8H_{16} may be neutral *hypho*-boranes, but the only confirmed examples contain donors, such as $B_6H_{10}(PMe_3)_2$ (a *hypho*-derivative of $[B_9H_9]^{2-}$). Many complex boranes have also been identified in which two simple polyhedra (open or closed) are joined. These are known as *conjuncto*-boranes , and include such species as $B_{10}H_{16}$ in which two B_5H_9 groups are joined by a $B-B$ bond (Greenwood & Earnshaw (1984), p. 173).

9.6.3 Molecular orbitals for $[B_6H_6]^{2-}$

The background to Wade's classification of the boranes is found in the molecular orbitals of $[B_6H_6]^{2-}$. Each boron can be regarded as sp hybridized, with one sp orbital forming an electron-pair bond to the *exo*-hydrogen (Fig. 9.14). The B_6 cluster has 18 atomic orbitals: 6 *radial* sp hybrids pointing from each boron to the centre of the cluster, and 12 *tangential* orbitals comprising the unhybridized p orbitals on each boron. These combine to give 18 molecular orbitals of which 7 are bonding (Fig. 9.15a). The most stable of these (the a_{1g} orbital) is formed by the in-phase combination of all the radial orbitals (Fig. 9.15b); the others involve the tangential p orbitals, combined head-to-tail (t_{2g}) or side-by-side (t_{1u}) around one B_4 plane. The existence of these *seven* bonding orbitals explains why the B_6H_6 cluster is stable as the dianion, $[B_6H_6]^{2-}$.

Qualitatively, we can also understand why all the *closo*-boranes are dianions. Each has its boron atoms hybridized in the same way as in the B_6 cluster. Each

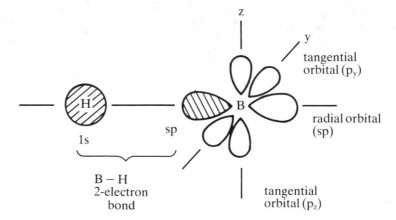

Fig. 9.14 — Hybridization of B-atom in a closo-borane, such as $[B_6H_6]^{2-}$.

† A simple method of calculating the number of skeletal electrons is the following: convert the species to a neutral borane by adding H^+, replacing CH by BH_2, etc. Then, the number of skeletal electrons is the same as the total number of atoms, e.g. 14 for B_4H_{10}.

(a)   Mo's of B₆H₆ cluster

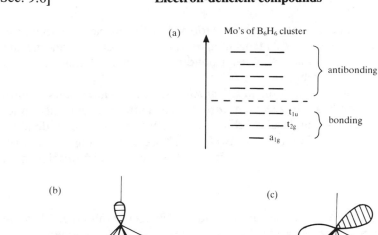

(b)   (c)

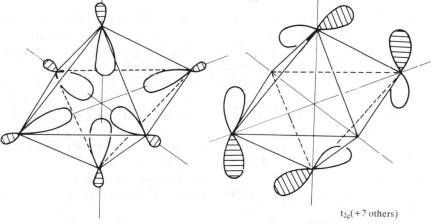

a₁g (radial *sp* hybrid AO's)

t₂g(+2 others)

(tangential *p* AO's)

(d)

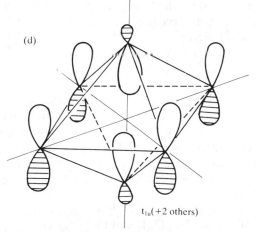

t₁u(+2 others)

(radial and tangential AO's)

Fig. 9.15 — Molecular orbitals of a B₆ octahedron (a) energy diagram (b) a1g orbital (c) one t₂g orbital; the two others are similar, relating to the other two B₄ planes (d) one t₁u orbital. (b)-(d) reproduced with permission from Wade, K., *Adv. Inorg. Chem. Radiochem.*, **18**, 1 (1976).

therefore has *one* low-energy molecular orbital formed by the in-phase combination of the n radial atomic orbitals. The $2n$ tangential orbitals will produce n bonding and n antibonding orbitals, giving a total of $n + 1$ bonding molecular orbitals requiring $n + 1$ pairs of electrons for stability.

It is more difficult to show that the *nido-*, *arachno-*, and *hypho*-clusters require the same number of bonding orbitals as the *closo*-boranes, after the removal of one, two, or three vertices. The molecular orbitals can be worked out for individual examples, such as $[B_5H_5]^{4-}$ (formed from $[B_6H_6]^{2-}$), because of its high symmetry. We deduce from the experimental observations that similar relationships hold generally.

9.6.4 Wade's rules

The relationships between the boranes are summarized in a set of rules that allow the structures of the simple boranes to be predicted; they apply to the components of *conjuncto*-boranes, but not to the complete molecules. These rules were originally formulated by K. Wade and are known as *Wade's rules* or the *skeletal electron-counting rules*. The rules (slightly modified from their original form) are as follows:

1. The structures of the boranes and their derivatives are based on polyhedra with triangular faces.
2. The skeletal atoms occupy all (*closo*), all but one (*nido*), all but two (*arachno*), or all but three (*hypho*) vertices of the appropriate polyhedron.
3. Each of the a skeletal atoms has a hydrogen atom (or some other singly-bonded ligand), attached by a bond pointing outwards from the centre of the polyhedron (the *exo* atom).
4. Apart from the electron pair in the *exo*-bond, the remaining valence shell electrons are the skeletal electrons (b pairs).
5. Each skeletal atom provides three atomic orbitals for skeletal bonding.
6. An n-atom parent closed polyhedron generates $n + 1$ skeletal bonding orbitals.
7. The boranes adopt a *closo*-structure if $b = (a + 1)$, *nido* if $b = (a + 2)$, *arachno* if $b = (a + 3)$, *hypho* if $b = (a + 4)$.

9.6.5 Metal clusters

One of the most remarkable links in modern inorganic chemistry has been made between the clusters of boron atoms in boranes and clusters of metal atoms in metal carbonyls. It has been found that skeletal electron counting can also be used to explain the structures of these carbonyls. This area of chemistry has been explored in greater depth than can be covered here, so only a few examples are presented, to show the similarities between the two types of cluster. (For further details, see Wade 1976, McPartlin 1984).

It is at first sight difficult to relate $C_2B_4H_6$ and $[Os_6(CO)_{18}]^{2-}$ (Fig. 9.16a), although both contain octahedra, C_2B_4 and Os_6 respectively. In reality, the difference between these two clusters is only the difference between any main group element and a transition metal. Boron has four valence orbitals (s + p), but osmium has nine (s + p + d). Compounds of both are stable when all their orbitals are involved in bonding (cf. section 8.3.2). In boron hydride clusters, this means that one

(a)   (b)

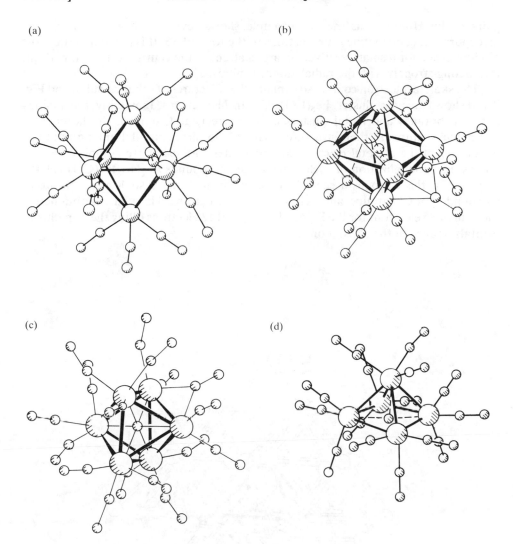

(c)   (d)

Fig. 9.16 — Metal clusters (a) $[Os_6(CO)_{18}]^{2-}$, (b) $Rh_6(CO)_{16}$, (c) $Ru_6(CO)_{17}C$, (d) $Fe_5(CO)_{15}C$. Redrawn after Wade (1976).

orbital contains a B-H electron pair and *three* contribute to the cluster orbitals, as just discussed. In osmium clusters, therefore, six orbitals form electron pair bonds (or contain lone pairs) and again *three* are involved with the cluster.

We have seen that an octahedral cluster is stable with 14 skeletal electrons. $[B_6H_6]^{2-}$ therefore has $14 + 6 \times 2 = 26$ electrons in total, while a similar Os_6 cluster will need $14 + 6 \times 12 = 86$ electrons. $[Os_6(CO)_{18}]^{2-}$ has in all $2 + 6 \times 8 + 18 \times 2 = 86$ valence electrons, exactly as needed for a *closo*-cluster. The same electron count can be achieved in other ways, e.g. in $Rh_6(CO)_{16}$ and $Ru_6(CO)_{17}C$ (Figs 9.16b,c). The central cluster is not affected by the replacement of CO ligands by lone pairs or by H-atoms; as the last example shows, even a carbon atom can be incorporated, contributing four electrons to the total of 86. It is located in the most obvious place for a small positively-charged atom, in the centre of the octahedron, interacting strongly with the radial atomic orbitals.

The skeletal electron count also explains the structures of other clusters, and Fig. 9.16d shows the *nido*-cluster, $Fe_5(CO)_{15}C$ with 74 electrons. Inevitably, the analogy between boranes and metal clusters is not perfect, and some of the latter have distinctive features in their structures that find no direct parallel among boranes. Extension of the skeletal electron counting rules to these more complex systems allows the rationalization of such remarkable compounds as $[Ni_{38}Pt_6(CO)_{48}H_2]^{4-}$ (Fig. 9.17); this contains a central Pt_6 octahedron within a shell of 38 Ni atoms surrounded by CO groups, and has a total of 542 electrons. This figure is obtained as the sum of 86 electrons for the Pt_6 octahedron and 12 electrons filling the non-cluster orbitals of each of the 38 Ni atoms.

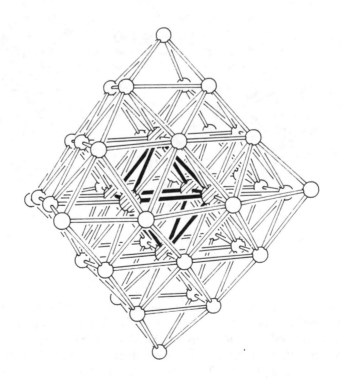

Fig. 9.17 — The $[Ni_{38}Pt_6(CO)_{48}H_{6-n}]^{n-}$ metal cluster, whose formula can be rationalized by an extension of Wade's rules. Reproduced with permission from Longoni,G. *et al.* (1985) *Angew. Chem. Int. Ed.* **24**, 697.

9.7 DELOCALIZED ELECTRONS AND AROMATICITY

When double bonds in an unsaturated organic molecule are formed by adjacent atoms, e.g.. atoms 2 and 3 in butadiene (III), the p orbitals on these atoms interact. The π-molecular orbitals are therefore not restricted to the 1–2 and 3–4 bonds, but cover the whole molecule. Such adjacent double bonds are called *conjugated* and the π-electrons are *delocalized* over the conjugated system. The results of conjugation are examined in this section, particularly the remarkable effects when double bonds are conjugated around a ring, as in benzene.

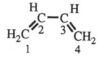

III Butadiene

Two qualifications should be noted to this general criterion for conjugated double bonds. For the best overlap between the p orbitals, they must be parallel, i.e. the conjugated section of the molecule must be planar. The overlap between non-parallel orbitals falls off as $\cos(\theta)$, where θ is the angle between them, so a modest amount of non-planarity has little effect. In Fig. 9.18., overlap is a maximum for geometry (a) and good for (b), but zero for (c). Secondly, two double bonds formed

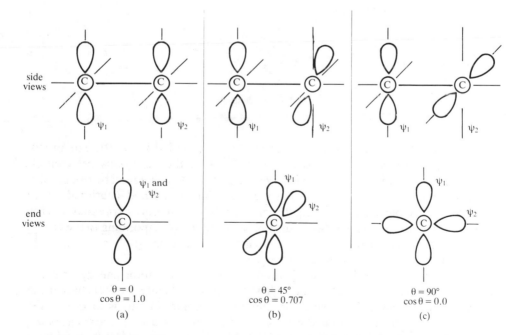

Fig. 9.18 — Conjugation of π-orbitals: side-on and end-on views of (a) parallell, (b) intermediate, (c) perpendicular orbitals.

from the *same* atom are not conjugated. In (IV), the 1−2 bond uses the p_x orbitals on atoms 1 and 2, while the 2−3 bond uses the p_y orbitals. As these orbitals are mutually perpendicular, their overlap is zero $(\cos(90) = 0)$ and the double bonds have no interaction. Such adjacent double bonds are termed *cumulated* rather than conjugated, and (IV) is the simplest *cumulene*.

$$H_2C=C=CH_2$$
$$1 \quad 2 \quad 3$$

IV Allene

9.7.1 Molecular orbitals in conjugated systems

The effect of conjugation on the molecular orbitals can be seen by comparing the localized orbitals in C_2H_4 with the delocalized orbitals in butadiene (III) and hexatriene (V) (Fig. 9.19). They have respectively four and six π-symmetry p orbitals, and therefore four and six molecular orbitals. In each, the most stable molecular orbital obviously has all the atomic orbitals overlapping in phase, and the higher energy orbitals gradually introduce more out-of-phase overlaps. The three molecules have respectively one, two, and three bonding orbitals, which are filled by the two, four, or six π-electrons.

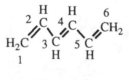

V Hexatriene

The energies of these orbitals can also be calculated (Fig. 9.20). For precise results, sophisticated computation is required, taking the repulsions between the individual electrons into account. However, approximate values can be obtained by a simple and elegant approach, known as the Hückel method (outlined in the appendix to this chapter). Following this method the energies are expressed in terms of two parameters; α is the energy of an atomic p orbital (corresponding to the energy of a non-bonding electron) and β, known as the 'resonance integral', is related to the interaction between two individual p orbitals.

The most obvious effect of conjugation is on the promotion energy of the π-electrons. In C_2H_4, ψ_b (HOMO, the highest occupied molecular orbital) has energy $\alpha-\beta$ and ψ_{ab} (LUMO, the lowest unoccupied molecular orbital) is at $\alpha + \beta$. The excitation of an electron from the lower to the higher π orbital (conventionally labelled $\pi \rightarrow \pi^*$) produces absorption at 7.62 eV. As the conjugated system grows in size, so the spacing of the levels decreases, giving HOMO–LUMO transition

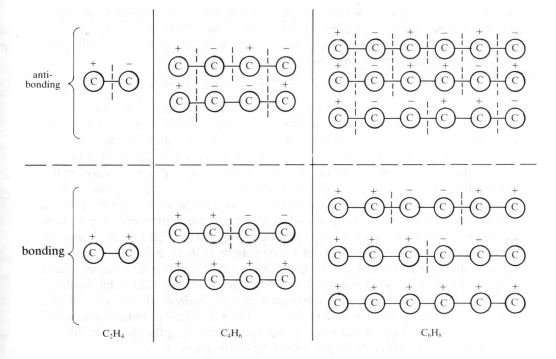

Fig. 9.19 — Molecular orbitals in C_2H_4, C_4H_6 and C_6H_8. The p orbital combinations for each molecular orbital are viewed perpendicular to the plane of the molecule. The dashed line separates bonding and anti-bonding orbitals.

$$\psi_6 \longrightarrow \alpha + 1.802\beta$$

$$\psi_4 \longrightarrow \alpha + 1.618\beta$$

$$\psi_5 \longrightarrow \alpha + 1.247\beta$$

$$\psi_{ab} \longrightarrow \alpha + \beta$$

$$\psi_3 \longrightarrow \alpha + 0.618\beta$$

$$\psi_4 \longrightarrow \alpha + 0.445\beta$$

$$\alpha \ \underline{\quad\quad} \ \underline{\quad\quad} \ \underline{\quad\quad} \ \underline{\quad\quad} \ \underline{\quad\quad} \ \underline{\quad} \ \alpha$$

$$\psi_2 \overset{\uparrow\downarrow}{\underline{\quad}} \alpha - 0.618\beta$$

$$\psi_3 \overset{\uparrow\downarrow}{\underline{\quad}} \alpha - 0.445\beta$$

$$\psi_b \overset{\uparrow\downarrow}{\underline{\quad}} \alpha - \beta$$

$$\psi_2 \overset{\uparrow\downarrow}{\underline{\quad}} \alpha - 1.247\beta$$

$$\psi_1 \overset{\uparrow\downarrow}{\underline{\quad}} \alpha - 1.618\beta$$

$$\psi_1 \overset{\uparrow\downarrow}{\underline{\quad}} \alpha - 1.802\beta$$

C_2H_4 C_4H_6 C_6H_8

Fig. 9.20 — Energies of molecular orbitals for C_2H_4, C_4H_6, and C_6H_8.

energies of 5.71 and 4.93 eV in butadiene and hexatriene. These values agree well with the simple theory, if a value of 2.62 eV (253 kJ mol^{-1}) is used for β.

Conjugation also increases the stability of the molecule as a whole. Thus, the total energy of the six π-electrons in hexatriene is

$$6\alpha - (2 \times 1.802 - 2 \times 1.247 - 2 \times 0.445)\beta = 6\alpha - 6.988\beta.$$

This can be compared to $(6\alpha - 6\beta)$ for the energy of three isolated ethene-like double bonds. The difference, the extra stability of the conjugated system, is known as its *delocalization energy*. It can be compared to an experimental quantity, the *resonance energy*: the difference between the actual heat of formation of the compound and that predicted by adding the bond energies for the individual bonds, ignoring the conjugation. The resonance energy can also be estimated by comparing the heats of hydrogenation of conjugated molecules and similar compounds with non-conjugated double bonds (more accurate because fewer measured values are needed). For *trans*-hexatriene, the difference between its heat of hydrogenation (332 kJ mol^{-1}) and the sum of the heats of hydrogenation of 1.5-hexadiene (248 kJ mol^{-1}) and 3-hexene (113kJ mol^{-1}) is 29 kJ mol^{-1}. The great discrepancy between the corresponding value of β (29.3 kJ mol^{-1}) and that calculated above from the ultraviolet spectrum (253 kJ mol^{-1}) arises because the second relates to the individual energy levels, the first to the system as a whole (where the approximations in the energy calculations have more effect). The second value is important in showing that linear molecules like hexatriene are only modestly stabilised by conjugation.

9.7.2   Aromatic molecules: benzene

Joining the two ends of hexatriene to give benzene (VI) has a remarkable effect on the molecule's stability and properties. This appears most obviously in comparing its chemistry to that of simple or conjugated unsaturated molecules. Thus, benzene cannot easily be oxidized (even by HNO_3) and is thermally stable. When it does react, say with Br_2 in the presence of pyridine, the product is not an addition product as with unsaturated molecules, but a substituted benzene. All these properties are shared by a large number of molecules which are more or less closely related chemically to benzene. These molecules are labelled *aromatic*. Although the original definition of aromatic character was based on chemical properties, physical methods have more recently been found to identify it. These are discussed in section 9.7.3.

VI   Benzene

The π-molecular orbitals for benzene are not very different from those for hexatriene. Joining the ends of the chain stabilizes orbitals with end-lobes of the same phase (ψ_1, ψ_3, ψ_5 in hexatriene, Fig. 9.19) by introducing an extra bonding overlap; orbitals with out-of-phase end-lobes are destabilized. The result (Fig. 9.21)

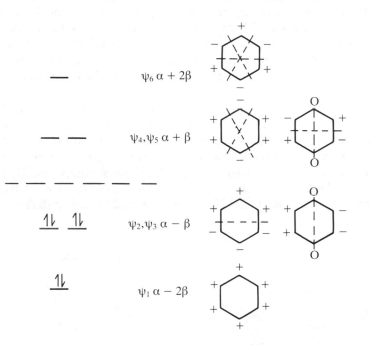

Fig. 9.21 — Molecular orbitals of benzene (a) energies, (b) geometry, indicating the phases of the orbital lobes on the upper side of the ring.

is that pairs of orbitals become degenerate: ψ_2, ψ_3, and ψ_4, ψ_5. As with hexatriene, the three bonding orbitals are filled and the total π-bond energy is $6\alpha - 8\beta$. Thus, the delocalization energy is 2β. Much effort has been devoted to experimental and theoretical determination of the resonance energy to compare with this value; the main problem is that of finding suitable model compounds. The currently accepted value is 83.7 kJ mol^{-1}, giving a value of 41.9 kJ mol^{-1} for β (Garrett, 1979, vol. I, p.232). However, this gives an underestimate of the thermodynamic stability of benzene. Relative to the double bonds in cyclohexene (taken three times), benzene is 150.4 kJ mol^{-1} more stable. It is this difference that underlies the chemical differences between aromatic molecules and olefins. Benzene is also very much more stable than hexatriene and similar conjugated but non-cyclic olefins. The simple calculation of π-bond energy suggests that β should be the same in both molecules, and shows a difference of only 1.01β between them. However, the limitations of the simple molecular orbital treatment and its neglect of the interaction between electrons mean that this is misleading. Additional stabilization of hexatriene would require the development of positive and negative charges at the ends of the molecules, unfavourable on electrostatic grounds; in cyclic molecules like benzene, these charges remain in close proximity.

9.7.2.1 *The surprising structure of benzene*

The structure of benzene has been determined experimentally with great precision. It is a regular planar hexagon, with all C−C distances 139.7 pm and all C−H distances 108.4 pm. This is expected from the simple molecular orbital treatment. If the total π-electron density is determined, by adding the contributions of each occupied molecular orbital, it is found to have equal values at each atom and in each bond. The C−C bond distances correspond to a bond order of approximately 1.5. As a π-bonded ring of sp^2 carbon atoms, the structure of benzene is exactly as would be expected.

However, it is also possible to consider benzene from the viewpoint of section 9.6, as a boron hydride with all the boron atoms replaced by carbon. Excluding the *exo*-bonds to hydrogen, it has 18 skeletal electrons. As it contains six cluster atoms, it should be an *arachno*-derivative of an 8-vertex polyhedron, with two vacant vertices. It is in fact isoelectronic with $[B_6H_6]^{6-}$ and the known compound B_6H_{12}. The structure of the latter (based on n.m.r. evidence) appears to be related to a dodecahedral borane cluster (Fig. 9.22a). Benzene, of course, is different.

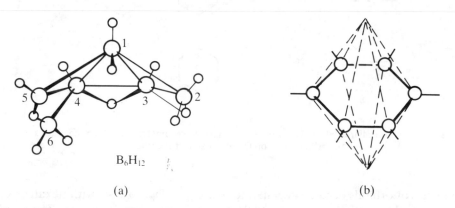

B_6H_{12}

(a)   (b)

Fig. 9.22 — (a) Inferred structure of B_6H_{12}, (b) Benzene viewed as a hexagonal bipyramid with two vacant positions. (a) reproduced with permission from Greenwood, N. N. & Earnshaw, A. (1984) *Chemistry of the elements*, Pergamon Press, Oxford, p.174.

These contrasting views can be reconciled by regard benzene as derived from a different 8-vertex polyhedron, a hexagonal bipyramid rather than a dodecahedron, with two *trans* rather than two *cis* vertices removed (Fig. 9.22b). Clearly, along the isoelectronic series

$$B_6H_{12} \rightarrow CB_5H_{11} \rightarrow \dots \rightarrow C_5BH_7 \rightarrow C_6H_6$$

the most stable structure changes. At present this change cannot be investigated more closely. Apart from the end members, only the C_5B example has been synthesised, and that only as a substituted anion $[H_5C_5BPh]^-$; it apparently has a planar hexagonal structure.

It is perhaps fortunate for the progress of organic chemistry that benzene has a hexagonal structure. Rationalizing its bonding, reactions and substitution patterns would have been far more difficult if it adopted the dodecahedral cluster structure.

9.7.3 Aromatic properties

The aromatic character of a compound was originally identified from its chemical
stability, in particular its ability to undergo substitution reactions rather than
addition across double bonds. However, it is impossible to produce a quantitative
measure of aromaticity from a compound's reactivity. Furthermore, the route a
reaction takes, whether to addition or substitution, depends on the reaction
intermediate as well as on the stability of the reactant and product.

Physical techniques have therefore become the main method of measuring
aromatic character. One obvious approach is to determine the resonance energy.
The possession of substantial resonance energy seems to characterize aromatic
compounds, but its calculation requires numerous assumptions which have been the
subject of considerable disagreement — as indicated by the range of 55 to
480kJmol^{-1} proposed for the resonance energy of benzene. The absence of alternation in length between the apparently single and double bonds is a useful guide but
has to be applied with caution; naphthalene (VII) is certainly aromatic but has three
different C−C distances.

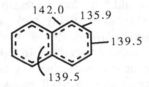

142.0   135.9

139.5

139.5

VII Naphthalene

The essential feature of an aromatic compound is now considered to be its
possession of a cyclic delocalised π-system. The best tests for aromatic character
detect this directly. If a magnetic field is applied to an aromatic compound, a current
is induced in the cyclic π-system (Fig. 9.23). This current, known as a *ring current*,
produces a magnetic field which opposes the original field within the ring and (to a
lesser extent) reinforces this field outside the ring.

This induced magnetic field can readily be observed through its influence on the

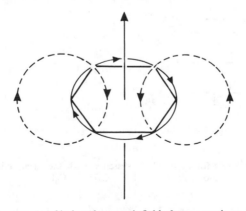

Fig. 9.23—The ring current and induced magnetic field of an aromatic compound in a magnetic
field.

chemical shifts in the ^1H n.m.r. spectra of protons near the ring. Those on the outside of the ring move to low field, e.g. δ 7.27 for the benzene protons, compared to about 5.5 for normal olefinic protons; the chemical shifts for protons within the ring move to high field. These are known as *diatropic* shifts. Benzene has no protons within the ring but other aromatic systems do (see p. 251), while benzene will influence the chemical shifts of protons in other molecules that are positioned near the molecular axis.†

9.7.4 Other aromatic compounds

The remarkable properties of benzene have led to the search for aromatic properties in other C_nH_n rings. In predicting their behaviour, the first step is to calculate the energy levels. It has been found that application of the Hückel technique to rings is rather simple. The ring is drawn inside a circle of radius 2β with one vertex pointing down (shown in Fig. 9.24 for C_5H_5, C_3H_3, and C_8H_8 as examples). The energy levels are then obtained by projecting the vertices onto an energy scale.

Knowing these energy levels, the properties of the various rings can be predicted. A neutral C_5H_5 molecule would have five π-electrons, leaving one of its bonding orbitals half-occupied. It therefore readily picks up an extra electron to give $[C_5H_5]^-$, the cyclopentadienide anion. This is stable, though fairly reactive, and shows ring current shifts in its n.m.r. spectrum. In C_3H_3, the pattern of energy levels (Fig. 9.24b) means that the neutral molecule has its unpaired electron in an *anti-bonding* orbital. For stability it therefore needs to lose this electron to give $[C_3H_3]^+$,

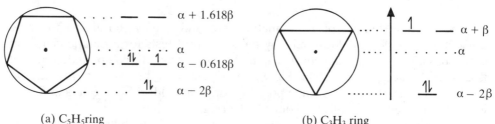

(a) C_5H_5 ring   (b) C_3H_3 ring

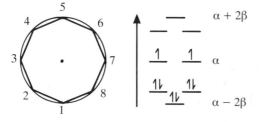

(C) C_8H_8 ring

Fig. 9.24 — Energy levels for cyclic C_nH_n systems, with the occupied levels for neutral molecules (a) C_5H_5 (b) C_3H_3 (c) C_8H_8.

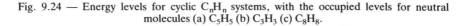

† The induced magnetic field can also be detected through its effect on the diamagnetic behaviour of an aromatic compound, though this is more difficult to observe; see Garrett (1986).

the cyclopropenium cation; this has been prepared as its $[SbCl_6]^-$ salt which readily reacts with water to give cyclopropanol but is thermally stable. The C_7 and C_9 rings behave similarly to C_3 and C_5 respectively, giving aromatic ions, $[C_7H_7]^+$ and $[C_9H_9]^-$, as regular polygons.

The distinction between C_3H_3, C_5H_5, and C_6H_6 can be systematized by noting that they have respectively $4n - 1$, $4n + 1$ and $4n + 2\pi -$ electrons. Each attains a set of completely filled bonding π orbitals when they have $4n + 2$ π orbitals (2, 6, and 6 electrons respectively). This was first recognized by Hückel, and expressed as the Hückel rule: that $4n + 2$ π electrons are needed for an aromatic system.

Other potentially aromatic systems have been the target of much recent synthetic effort with interesting results. The neutral C_nH_n molecules that obey the Hückel rule have been made, from $C_{10}H_{10}$ to $C_{22}H_{22}$; they are called *annulenes*. The best known is [18]-annulene (VIII). It is not a regular [18]-gon, but some of its vertices point

VIII 18-Annulene

inwards to reduce the strain in the C−C−C angles.† As a result, the ^1H n.m.r. spectrum demonstrates its aromaticity directly, with signals for the two sets of protons at $\delta - 2.99$ (inner) and 9.28 (outer) (at $-70°C$). Its C−C bonds are nearly but not quite uniform, with twelve of 138.2 pm and six of 141.9 pm, and it will undergo aromatic substitution. The other members of the series also show ring currents, though they are chemically less stable. [10]-annulene can be prepared only if two inward-pointing C−H groups are replaced by a bridging group, e.g. −CH₂− (IX). Otherwise the two C-H groups collide, forcing the ring to become non-planar and non-aromatic.

IX 1,6-Methano-[10]-annulene

---

† The calculated energy levels for [18]-annulene are not affected by this conformational change if the simple Hückel method is used, because this considers only nearest-neighbour interactions. More sophisticated calculations take the ring geometry into account, though this has little effect on the energy levels.

Hückel's rule applies directly only to monocyclic systems, though in principle polycyclic molecules can form π-molecular orbitals around their perimeter. More often, parts of the molecule behave as semi-independent aromatic systems. Thus, naphthalene (VII) behaves more like a derivative of benzene than as a bridged [10]-annulene.

9.7.5 Anti-aromaticity

We have seen that molecules with $4n - 1$ or $4n + 1$ π-electrons readily lose or gain an electron to achieve aromaticity. What about $4n$ π-electrons? Fig. 9.24c shows that C_8H_8, cyclo-octatetraene, has two unpaired electrons in degenerate non-bonding π-orbitals. Such a configuration is known as anti-aromatic. However, it is stabilised by distortion, to remove the degeneracy of the half-filled orbitals. This has a dramatic effect, converting the molecule into a non-planar tetraene (X). Its double bonds are almost perpendicular to each other, so the p orbital overlap is almost zero. C_8H_8 can become aromatic by either gaining or losing two electrons. $[C_8H_8]^{2-}$ and $[C_8H_8]^{2+}$ are both known, though the former is much more stable; this indicates a deficiency in the Hückel calculation of their energy levels, which would predict the opposite. Both are regular octagons. The smallest anti-aromatic ring, C_4H_4, can be stabilised only if it has bulky substituents instead of hydrogen; it then distorts to take up a structure with two double and two single bonds, rather than equal C–C distances. It also forms transition-metal complexes, such as $(C_4H_4)Mo(CO)_4$. In these, the π-orbitals of the cyclobutadiene interact with the d orbitals of the metal atom.

X cyclotatretraene

With larger rings, the destabilization of a $4n$ anti-aromatic system is less severe, just as the aromatic stabilization of $4n + 2$ electrons is reduced. Several $4n$-annulenes have been made, of which [16]-annulene is the best characterized (XI). Its crystal structure shows that it is nearly planar with alternating bond lengths (145.4 and 133.3 pm). A 16-membered ring with equal C–C bonds has a similar arrangement of energy levels to C_8H_8, with two singly occupied non-bonding orbitals. The observed alternation changes the energies of this pair of orbitals so that one is rather lower and

XI 16-Annulene

contains both electrons. However, the arrangement with two unpaired electrons is not very much higher in energy. A magnetic field stabilizes the parallel-spin configuration, so [16]-annulene shows slight paramagnetism with an induced magnetic field reinforcing the external field (the reverse of Fig. 9.23). This stabilization is not sufficient for the molecule as a whole to be paramagnetic, but it reduces the overall diamagnetism (the reverse of the increased diamagnetism observed with aromatic systems). The effect of the induced magnetic field appears in the ^{1}H n.m.r. spectrum, where the inner protons experience a *higher* magnetic field than the outer ones; they have δ 10.43, while the outer ones shift to δ 5.4 (at − 110°C); these are *paratropic* rather than diatropic shifts. Like C_8H_8, [16]-annulene will give both a dication and a dianion, and both show *aromatic* (diatropic) rather than *anti-aromatic* (paratropic) ^{1}H n.m.r. shifts.

9.7.6 Hetero-aromatics

As has just been seen, aromatic $4n + 2$ π-electron systems can be created by adding electrons to non-aromatic systems. The same result can be obtained by replacing a carbon atom with an atom having more electrons. The most notable example is pyrrole (XII). The structural diagram suggests that it should resemble butadiene, but

XII Pyrrole

experimentally its chemical properties are closer to those of benzene. The key to this behaviour is that it is a completely planar molecule, with sp^2 rather than sp^3-hybridized nitrogen. Redrawing the structure (Fig. 9.25) shows that its five π orbitals are conjugated, with molecular orbitals like those of C_5H_5 (Fig. 9.24a). The nitrogen lone pair gives it 6 π-electrons and the same aromatic configuration as $[C_5H_5]^-$.

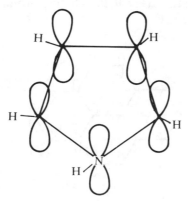

Fig. 9.25 — π-orbitals in pyrrole.

Many other examples of small hetero-aromatic systems are known, and the same principles also apply to larger rings. Thus, (XIII) has an aromatic 8-membered ring, corresponding to $[C_8H_8]^{2-}$. It is also possible to make further isoelectronic substitutions in such rings, producing remarkable aromatic systems (Morris & Rees 1986). In (XIV), each sp^2 carbon contributes one π-electron, while each sulphur provides a pair of electrons in a p orbital. The nitrogen atoms have a lone pair and *one* π-electron (in contrast to pyrrole where the N−H group provides two π-electrons). The ring has a total of 10 π-electrons and is aromatic. The seven-membered ring (XV) has the same number of electrons, while 14 π-electrons are present in the bicyclic ten-atom ring of (XVI). Here, as with pyrrole, the inclusion of electron-rich N and S-atoms reduces the number of of ring atoms needed for a given number of π-electrons, compared to aromatics containing only carbon atoms.

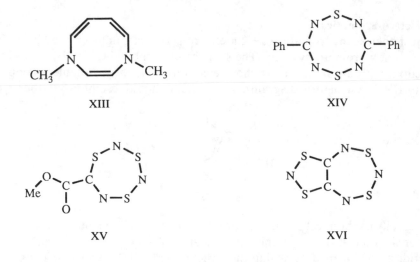

XIII   XIV

XV   XVI

9.7.7 Graphite
Graphite is the ultimate example of a compound with a conjugated π-system. In its individual layers of carbon atoms (Fig. 9.26), all the p orbitals overlap as in benzene.

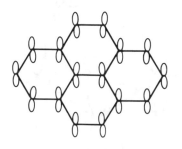

Fig. 9.26 — π-orbitals in part of a graphite layer.

Because graphite has infinitely extended layers, these orbitals produce a bonding and an anti-bonding *band* of molecular orbitals. The bonding band corresponds to the three bonding orbitals of benzene and contains just enough orbitals to hold one electron per carbon atom, i.e. all the π-electrons. The anti-bonding band just touches this at zero energy (Fig. 9.27); this behaviour is known as *semi-metallic.*† It leads to

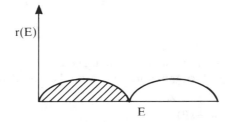

Fig. 9.27 — Band structure of graphite.

graphite being a conductor, because electrons are thermally promoted into the upper band. This conductivity takes place along the planes, but there is very little electron interaction between the layers, and graphite is an insulator in this direction. The effect on this conductivity of inserting atoms between the planes of graphite is examined in the final chapter (section 11.5.2).

9.8   APPENDIX: THE HÜCKEL CALCULATION OF MOLECULAR ORBITAL ENERGY LEVELS FOR C_nH_n SPECIES

The Hückel method assumes that only interactions between directly bonded atoms (magnitude β) are significant; all other interactions are taken to be zero. A determinant (the *secular determinant*) is constructed, reflecting the molecular geometry. This has $-\varepsilon$ (the energy of a molecular orbital relative to the non-bonding level α) for each diagonal term. The $i-j$ off-diagonal term is β if atoms i and j are bonded, 0 otherwise. Two examples (allyl, XVII; bicyclobutene, XVIII) illustrate this:

XVII Allyl   XVIII Bicyclobutene

† In contrast to metals where the bands overlap, and semiconductors (section 11.1) which have a gap between the bonding and anti-bonding molecular orbitals.

The secular determinants are

Atom	1	2	3
1	$-\varepsilon$	β	0
2	β	$-\varepsilon$	β
3	0	β	$-\varepsilon$

$= 0$

Atom	1	2	3	4
1	$-\varepsilon$	β	β	β
2	β	$-\varepsilon$	β	0
3	β	β	$-\varepsilon$	β
4	β	0	β	$-\varepsilon$

$= 0$

Equating the determinant to zero and expanding gives a polynomial whose solutions (roots) are the energy levels of the molecular orbitals. For allyl, the polynomial is

$$- \varepsilon(\varepsilon^2 - \beta^2) - \beta(-\varepsilon\beta) = 0$$

with roots $\varepsilon = 0$ and $\varepsilon = \pm\sqrt{2}\beta$ (i.e. energies α and $\alpha \pm \sqrt{2}\beta$). For more complex equations, determining the roots of these polynomials is not trivial, and various techniques have been developed to assist this. Approximate values can be obtained by plotting the value of the polynomial as ε varies, knowing that ε must lie in the range $\pm 2\beta$ for non-branched rings and chains or $\pm 3\beta$ for branched ones. It is also helpful to know that $\Sigma\varepsilon = 0$.

The secular determinant also represents a series of linear simultaneous equations whose unknowns are the coefficients of the atomic orbitals on each atom, used to construct each molecular orbital.

$$\psi = c_1\psi_1 + c_2\psi_2 + c_3\psi_3 \ldots$$

Thus, for allyl the equations

$$- \varepsilon c_1 + \beta c_2 + 0 = 0$$
$$\beta c_1 - \varepsilon c_2 + \beta c_3 = 0$$
$$0 + \beta c_2 - \varepsilon c_3 = 0$$

apply to each of its three orbitals (with the appropriate values of ε). These equations give the ratios of the coefficients (c_2/c_1, c_3/c_1, etc.), and their absolute values can be obtained by normalizing (Box 7), using the extra equation:

$$c_1^2 + c_2^2 + c_3^2 = 1.0$$

These equations are solved with each of the values of ε obtained previously, to give the coefficients for the molecular orbital of that energy. For ψ_2 of allyl ($\varepsilon = 0$), the equations become

$$\beta c_2 = 0 \text{ and}$$
$$\beta c_1 + \beta c_3 = 0,$$

i.e. $c_1 = \sqrt{2}$, $c_3 = -\sqrt{2}$, $c_2 = 0$, the non-bonding orbital of allyl (XIX).

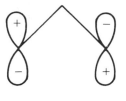

XIX

Completion of these equations for bicyclobutene is left to the reader; the results are given by Streitweiser (1961).

In a monocyclic C_nH_n system (as in Fig. 9.24), the energies are given by

$$\varepsilon_j = 2\beta\cos(2j\pi/n)$$

with j taking values ± 0, 1, 2 (up to $(n-1)/2$ for n odd or $+n/2$ (not $-n/2$) for n even).

FURTHER READING

d-Orbitals
Craig, D. P. & Magnusson, E. A. (1956) *J. Chem. Soc.* p. 4895.
Ebsworth, E. A. V. (1987) Shapes of simple silyl compounds in different phases. *Acc. Chem. Res.* **20**, 295.

F−O compounds
Spratley, R. D. & Pimentel, G. C. (1966) *J. Amer. Chem. Soc.*, **88** 2394.

Electron Deficient compounds and clusters
Wade, K. (1976) Structural and bonding patterns in cluster chemistry. *Advances in Inorganic chemistry and radiochemistry*, 18,1 (1976).
Greenwood, N. N. (1973) Boron. In: Bailar, J. C. *et al.* (eds) *Comprehensive inorganic chemistry*, Pergamon, Oxford, Vol. I.
Greenwood, N. N. & Earnshaw, A. (1984), *Chemistry of the elements*, Pergamon, Oxford.
Housecroft, C. E. (1989) *Boranes and metallaboranes*. Ellis Horwood, Chichester.
McPartlin, M. (1984) Electron-counting procedures in clusters of the iron triad. *Polyhedron*, **3**, 1279.

A romaticity
Garrett, P. J. (1986) *Aromaticity*, Wiley, New York.
Garrett, P. J. (1979) In: Barton, D., & Ollis, W. D. (eds.) *Comprehensive organic chemistry*, Pergamon, Oxford.

Hückel calculations

Salem, L. (1966) *Molecular orbital theory of conjugated systems*, Benjamin, New
    York.
Streitweiser, A. (1961) *Molecular orbital theory for organic chemists*, Wiley, New
    York.

Text references

Bartell & Garvin (1968) *J. Chem. Phys.* **48** 2466.
J. L. Morris & C. W. Rees, *Pure and applied chem.*, **58** 1.

10

Complex ionic compounds: remarkable properties

10.1 INTRODUCTION

This chapter looks further at ionic compounds, mainly metal oxides. The obvious distinction between them and the simple ionic compounds discussed in Part II is that their formulae are more complicated: $LiNbO_3$ instead of TiO_2, for example. The introduction of two or three different cations greatly increases the range of formulae and structure. Among these compounds some have extraordinary properties, mechanical, electrical, magnetic, and optical. These arise partly through the interactions between the cations, but the detailed crystal structures themselves are also important. The fundamental bonding remains unaltered by these interactions and is still well described by the ionic model.†

In one way, the structural complexities must not be overstated. The replacement of some ions by others becomes more feasible as the structure becomes more complicated, because there is a wider choice of coordination number and geometry. The replacements need not even match the original charge, as it is often possible to remove other anions or cations to balance the charge (as in ZrO_2+CaO, section 10.4). As a result, though an enormous variety of compounds can exist, they take up remarkably few structures. Among ternary ionic compounds, those with three types of ions, only three main formulae are found, ABX_3, ABX_4 and AB_2X_4, divided into fifteen structural families. Other formulae and structures are restricted to one or two combinations of ions (Muller & Roy 1974). Even rarer are distinct quaternary structures, that have specific sites for four different ions rather than just being variants of ternary or binary structures. No known structural families contain five different ion sites.

The remarkable properties discussed in this chapter illustrate a few of the areas

---

† By no means all the properties discussed here are exclusive to ionic compounds, as will be seen in Chapter 11.

that solid-state chemists and physicists are beginning to explore. Others are examined in Chapter 11, and for a fuller review, the sources listed under *Further reading* can be consulted.

10.2 COMPLEX STRUCTURES

The first problem both in describing and investigating the field of complex ionic compounds lies in the choice available. Despite the small number of structural families, simple permutation of cations within a given formula produces an immense number of possible compounds; in ABO_3 compounds, where the charges of A and B add up to +6, simply selecting two of 40 possible M^{3+} ions leads to 1600 different compounds. As will be seen, two compounds with this structure can have entirely different properties, so all these combinations may need to be studied. Attempts to identify what controls the occurrence of particular structures have had some success. An impressive example concerns the group of ternary oxides of formula AB_2O_4. Fig. 10.1 shows how the major structural families are correlated directly with the

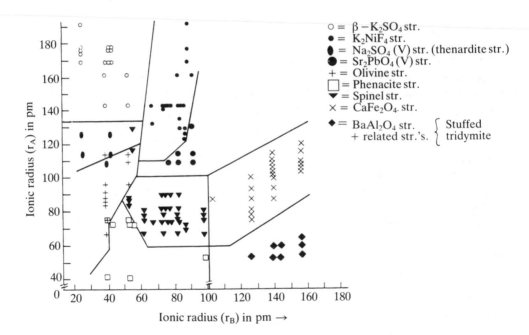

Fig. 10.1 — Sorting of AB_2X_4 compounds into structural families by the ionic radii of A and B. Each point on the plot represents at least one compound with the indicated structure. Reproduced with permission from Muller, O. & Roy, R. (1974) *The major ternary structural families*, Springer Verlag, Berlin, Fig. 6.

sizes of the ions A and B. Sometimes the formation of partial covalent bonding is also important in controlling the type of structure found, and the relative electronegativities of the compounds are significant; this is examined in section 11.3.

Two of the most characteristic ternary structures are described here. They reappear as individual examples with unusual properties later in the chapter. These are the spinel and the perovskite structures. Both are named after minerals, $MgAl_2O_4$, spinel, and $CaTiO_3$, perovskite, and both can be made with many different combinations of cations.

10.2.1 Spinels

In the spinels, the AB_2O_4 formula is usually built up from $A^{2+}+B^{3+}$ ions, though examples with $A^{4+}+B^{2+}$ and other charges are found (including some spinels with halide rather than chalcogenide ions). They show a very characteristic feature among complex oxide structures, that the O^{2-} ions are in a close-packed array (c.c.p. in spinel). The metal ions then occupy some of the octahedral and tetrahedral holes.

The cubic unit cell† contains 32 oxygen atoms and therefore 32 octahedral and 64 tetrahedral holes. A quarter of each of these are occupied in a regular pattern (Fig. 10.2). Because M^{2+} ions are generally larger than M^{3+}, we would expect M^{2+} in the

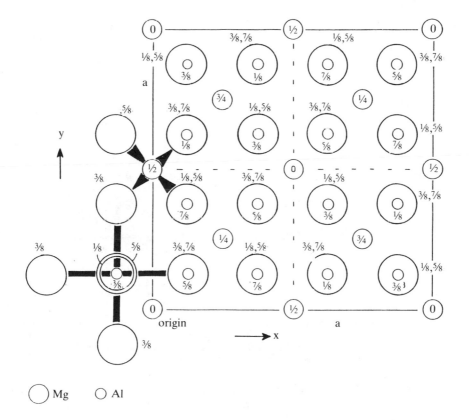

Fig. 10.2 — The spinel structure, seen in projection. The large circles are the close-packed oxygen ions. Reproduced with permission from West, A. R. (1984) *Solid state chemistry and its applications*, Wiley, Chichester, Fig. 16.9.

† Twice the length on each axis compared to the original close-packed cubic structure.

octahedral holes, M^{3+} in the tetrahedral holes. This happens for $MgAl_2O_4$, which is known as a *normal* spinel. However, many spinels fill the octahedral holes with M^{3+} and have mixed M^{2+}/M^{3+} in the tetrahedral holes. These are the *inverse* spinels, such as $NiFe_2O_4$, with Fe^{3+} octahedrally coordinated and Ni^{2+}/Fe^{3+} tetrahedral. Both the mixed valence oxides Fe_3O_4 ($Fe^{2+}Fe_2^{3+}O_4$) and Co_3O_4 are spinels, the first with the inverse and the second the normal structure.

Such apparently random selections of one or the other spinel type have led to many attempts to explain which metal ions give normal and which inverse spinels. One approach has been through crystal field stabilization energy (CFSE; section 8.2.2), to decide which of ions A or B gains most from an octahedral rather than a tetrahedral environment. This works fairly well, but is limited to the 70 spinels containing at least one ion with an incomplete d-shell. The map in Fig. 10.3 shows a

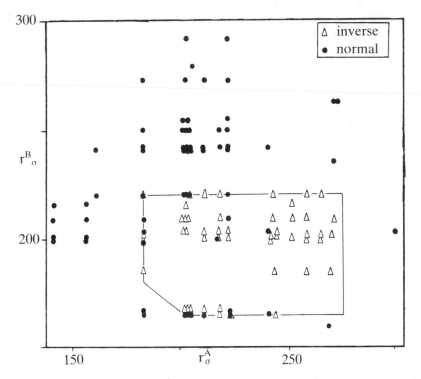

Fig. 10.3 — Sorting of AB_2X_4 spinels (X = chalcogenide) into normal and inverse types using the 'orbital' radii of A and B. Many of the points plotted represent more than one spinel. Reproduced with permission from Burdett, J. K., Price, G. D., & Price, S. L. (1982), *J. Amer. Chem. Soc.* **104**, 92, copyright American Chemical Society.

much more successful identification of the 40 inverse spinels, with only four errors among the 172 examples of spinels with O^{2-}, S^{2-}, and Se^{2-} ions. The spinels are sorted on radii, but using theoretically derived rather than experimental ionic radii.†

† These are 'orbital' or 'pseudo-potential' radii (see references in Burdett *et al.* 1982). The standard ionic radii used to define the overall structures of AB_2O_4 compounds (Fig. 10.1) do not discriminate among the spinels so successfully.

The CFSE is important, but only for spinels on the borderline between the normal and inverse types. We should note, though, that it is not yet possible to understand why the radii control the preference of the individual ions for octahedral or tetrahedral geometries.

10.2.2 Perovskites

The ABO_3 perovskite structure is most often found with $A^{2+}+B^{4+}$ ions, but can have $A^{+}+B^{5+}$ or $A^{3+}+B^{3+}$ as well. The coordination numbers of A and B are [12] and [6], and the structure is therefore stable when A is much larger than B. The following perovskites show this differentiation clearly: $KNbO_3$ (radii of the ions 152, 86 pm), $SrTiO_3$ (132, 81 pm), $LaMnO_3$ (117, 78 pm).†

The perovskite structure (Fig. 10.4) contains a cubic close-packed array of atoms

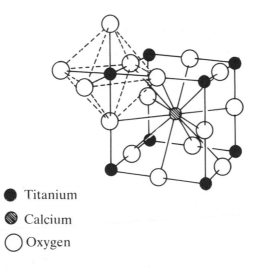

● Titanium

◍ Calcium

○ Oxygen

Fig. 10.4 — The perovskite structure, showing the 12-coordinate and 6-coordinate sites.

as in spinel, but this is not made up simply of the oxygen anions. Instead, it includes the A-ions as well, which are therefore 12-coordinated. The problem described in Chapter 4 for 12-coordination in ionic compounds (Fig. 4.3) does not arise, because enough O^{2-} ions are available to coordinate all the A-ions. The smaller B-ions occupy octahedral holes in this close-packed array; of the four holes associated with each group of AO_3 ions, they naturally choose the one which has only O^{2-} neighbours.

The perovskite structure is very sensitive to the precise sizes of the ions. As the A-ions become larger, the oxygen octahedra around each B-ion also expand. The structure may then distort to give the B-ions more favourable environments. Many perovskites therefore take up distorted structures, often transforming to undistorted

---

† For ions of more equal size, ABO_3 compounds usually adopt the ilmenite structure ($FeTiO_3$; radii Fe^{2+} 92.0 pm; Ti^{4+} 74.5 pm) with [6] coordination for both A and B.

cubic forms at higher temperatures when thermal vibrations change the effective sizes of the ions. As examples, $SrTiO_3$ is regular but $BaTiO_3$ distorted at room temperature (radii Sr^{2+} 132 pm, Ba^{2+} 49 pm).

10.3 BOND VALENCE CALCULATIONS

To a first approximation, the distances between ions in complex ionic compounds are predictable from standard ionic radii. However, precise structure determinations for less symmetrical compounds often reveal a range of distances. Remarkably, it has been found that each individual distance (atom i to atom j) can be equated to a 'bond valence', with the formula (10.1)

$$v_{ij} = \exp[(r_o - r_{ij})/b] \qquad (10.1)$$

very similar to that applied to covalent bonds (equation (5.1)).† For each ion (i), the sum of the bond valences

$$V_i = \Sigma v_{ij}$$

is then equal to the oxidation state of the ion. The empirical constants r_o and b have been obtained by fitting these equations to the distances in large numbers of structures (Brown & Altermatt 1985). As an example, Table 10.1 shows the

Table 10.1 — Bond valence calculations for $CaCrF_5$

F atom	(Coordination distances in pm)			
	Cr coordination by F	Bond valence	Ca coordination by F	Bond valence
F(1)	2×194.0	2×0.42	1×249.0	2×0.20
F(2)	2×191.8	2×0.46	2×229.1	2×0.29
			2×239.1	2×0.24
F(3)	2×184.8	2×0.65	2×224.4	2×0.33
Totals for Cr and ca		3.06		1.92
Bond valence sums for F	F(1): 2×Cr+1×Ca=1.02			
	F(2): 1×Cr+2×Ca=0.99			
	F(3): 1×Cr+1×Ca=0.98			

Source: Wu, K. K. & Brown, I. D. (1973).

calculation of the bond valences in $CaCrF_5$. This compound contains chains of distorted CrF_6 octahedra linked by Ca^{2+}. The table shows how the valences of Cr, Ca, and F correspond to their expected values. The calculation for F(3) reveals the

† When originally proposed, the equation $v_{ij}=(r_{ij}/r_o)^{-N}$ was used, with N and r_o as the empirical constants.

physical reality behind the bond valence sums. It is linked only to two cations, and as a result both these bonds are shorter and stronger than those involving the other fluorine atoms.

Bond valence calculations are particularly useful in identifying oxidation state. Thus, $FeTiO_3$ (ilmenite) might contain Fe^{2+} and Ti^{4+} and Fe^{3+} or Ti^{3+}. The bond valences (Fe 2.07; Ti 3.97) clearly confirm the first alternative. In V_4O_7 the mean bond valence is 3.5 as expected for a mixed valence oxide with this stoicheiometry; the individual valences vary between 3.28 and 3.63, showing that the proportion of V^{3+} and V^{4+} on each site varies.

The most remarkable feature of these calculations is the demonstration that the same basic relationship can be applied to the correlation of bond length and bond order in both covalent and ionic compounds. This foreshadows the link between these two extreme bonding types that is examined in Chapter 11.

10.4 'CERAMIC STEEL' AND 'FANTASY DIAMONDS'

The first example of an ionic compound with distinctive properties shows how minor modifications of crystal structure can remove the brittleness normally characteristic of ionic compounds. The basic material is zirconia, ZrO_2. Pure zirconia shows three phases, a monoclinic phase stable at room temperature, a tetragonal phase (1160°–2360°C), and cubic (2360°–2675°C, the melting point). Most MO_2 transition metal oxides have the [6:3] rutile structure (Fig. 4.8), but Zr is the largest M^{4+} ion and is just big enough to allow a higher coordination number. Thus, the cubic phase of ZrO_2 has the [8:4] CaF_2 structure (Fig. 4.8a; seen in side-view in Fig. 10.5a). The tetragonal phase involves a slight distortion; the arrows in Fig. 10.5b show how alternate sides of the cubes of O^{2-} ions are pushed in opposite directions. At low temperature, the Zr ion behaves as if it is too small for 8-coordination; presumably the thermal vibrations of the Zr-ion at high temperature give it a larger effective size. In the room-temperature monoclinic form, a major rearrangement of the structure takes place, with a reduction to 7-coordination for Zr. Fig. 10.5c shows how this distortion occurs. The O ion in the centre of the picture (and those related to it elsewhere in the unit cell) is now associated with just one Zr, rather than bridging between two ions.

Naturally, the reduction in coordination causes a large increase in the volume of the unit cell (5–10%). This is troublesome. It prevents zirconia being used as a refractory material despite its high melting point, because blocks or fabricated objects crack when the material changes phase. It has been found that including a proportion (15–20%) of a lower-valent metal oxide (e.g. CaO, MgO, Y_2O_3) causes the cubic phase to be stabilized, so zirconia is cubic from room temperature upwards. These other metals reduce the number of oxygen ions in the structure, allowing some of the Zr^{4+} ions to be seven-coordinate, even in the cubic CaF_2 structure. Intriguingly, this stabilized cubic zirconia has another use — as fake diamonds, with almost the same sparkle (resulting from its high refractive index) and hardness as the real thing, but easily grown as large crystals.

A much more interesting change takes place if the proportion of the added oxide is rather lower, 'partially stabilized' zirconia (PSZ) rather than 'fully stabilized'. Now the bulk of the material is cubic, but it contains tiny particles dispersed within it of the

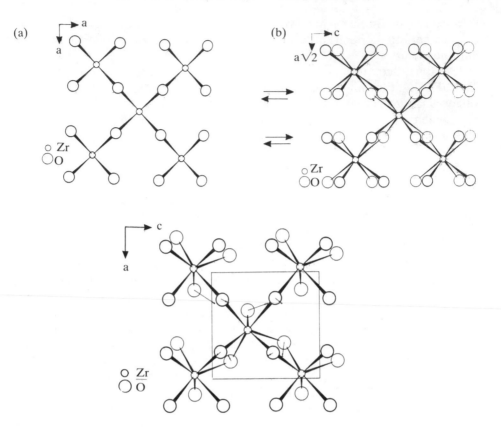

Fig. 10.5 — The structures of ZrO_2 (a) cubic form with regular cubes of oxygen atoms around each Zr, Zr–O distances 228.2 pm, (b) tetragonal form with slightly distorted cubes of oxygen atoms, four Zr–O distances of 206.5 pm and four of 245.5 pm, (c) monoclinic form with seven Zr–O distances in the range 204–226 pm and one of 377 pm. Thin lines indicate the shifts between (b) and (c). Source: Smith, D. K. & Newark, H. W. (1965) *Acta Cryst*, **18**, 983.

metastable tetragonal phase (100 nm or less in diameter). The effect of these on cracking is remarkable. Under stress, the tetragonal phase reverts to monoclinic. This occurs when a tiny crack meets a tetragonal particle. The resulting increase in volume partly fills the crack, and also removes the stress around it. The result is an enormous increase in the resistance of PSZ to brittle fracture, which has led to it being described as 'ceramic steel'. The strength of steel comes from minute inclusions of Fe_3C in the iron, which function in rather the same way under stress, so the name is not inappropriate.

10.5 SUPERIONIC CONDUCTORS

Ionic compounds are normally excellent insulators as solids, but when they melt, the ions are free to move through the liquid and they conduct electricity. A few ionic solids are exceptional in being good conductors (usually when heated) because some of their ions can move even in the crystal. These are known as 'superionic' or 'fast ion' conductors.[†] They are of increasing importance in relation to fuel cells and high temperature batteries which use, for example, the reaction of Na and S. The superionic conductor forms a membrane in the cell, allowing ions to pass from one side to the other, but preventing the reagents coming into direct contact.

10.5.1 Doped zirconia

The doping of ZrO_2 with lower-valent metal oxides (CaO, Y_2O_3, etc.) that stabilizes its cubic crystal structure (last section) produces a structure in which all the metal ion sites are occupied, but some of the oxygen ions are randomly missing. If this material is heated to 600°C, O^{2-} ions next to vacancies have enough thermal energy to move onto the vacant sites. Their own sites can then be filled, so that these ions become mobile in the crystal and conduct electricity (Fig. 10.6). This mobility allows

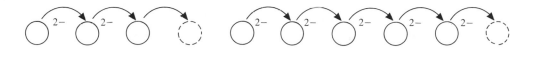

Fig. 10.6 — Schematic representation of the movement of O^{2-} ions in doped ZrO_2.

comparison of the oxygen concentration on each side of a crystal of doped zirconia (which functions as the membrane in an electrochemical cell). It is therefore applied as a high-temperature oxygen sensor, for example monitoring combustion efficiency in boiler flue gases.

10.5.2 Sodium β-alumina

A compound that was originally described as a form of Al_2O_3 and named β-alumina, has since been shown to require some M^+ ions for the structure to be stable. Its formula is variable, $M_2O.nAl_2O_3$, with $n=5$–11. The most important example is sodium β-alumina (typical formula $Na_2O.11Al_2O_3$, $NaAl_{11}O_{17}$) which is now used in Na/S high temperature batteries. Few other materials are inert at 300–400°C to molten sodium and sulphur, and also permeable to Na^+ ions, while not being metallic conductors (which would short-circuit the battery). The structure of sodium β-alumina shows how its conductivity arises (Fig. 10.7). It is derived from the spinel structure, $MgAl_2O_4$, with close-packed layers of O^{2-} ions, containing octahedral Al^{3+} ions, but with the Mg^{2+} sites vacant. Between each group of four layers (647 pm thick) is a gap (476 pm thick) which contains a layer of loosely packed Na^+ and O^{2-}

---

† Some transition metal oxides and similar compounds, such as VO, also conduct, but the current is carried by the electrons, i.e. is metallic in nature. This conductivity results from the overlap of the orbitals on different metal ions.

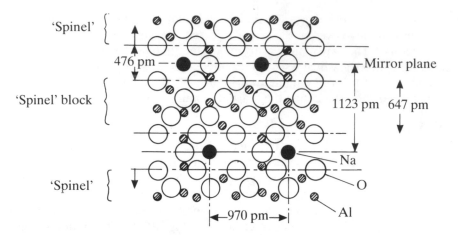

Fig. 10.7 — Sodium β-alumina structure. Reproduced with permission from Greenwood, N. N. & Earnshaw, A. (1984) *Chemistry of the elements*, Pergamon, London, Fig. 7.17.

ions. The Na^+ ions have a choice of several sites in this layer, as well as the ones they occupy in the ideal structure. As a result they move readily from site to site, giving significant conductivity even at room temperature, which rises with increasing temperature reaching about 10^{-1} ohm^{-1} cm^{-1} at 300°C.

Unlike ZrO_2, this conductivity is two-dimensional, strictly within the planes, and the material remains an insulator perpendicular to them. In batteries it is used as polycrystalline blocks to avoid the need to grow and orient single crystals, though these would be better conductors.

10.5.3   α-AgI

Silver iodide at room temperature is in its β-form with the wurtzite (ZnS) structure. At 146°C it transforms to the α-form. This has the highest ionic conductivity of any ionic solid, up to 1.0 ohm^{-1} cm^{-1}. It is higher even than in liquid AgI, so that the conductivity actually drops at the melting point. The structure of α-AgI is controlled by the very large I^- ions. They form a rigid body-centred cubic array, and the Ag^+ ions can wander over the surface of the I^- ions, avoiding only the points of contact between the latter. Fig. 10.8 gives an impression of the structure, showing the variety of positions through which each Ag^+ moves.

An important aim for studies of AgI has been to find ways of obtaining similar conductivity at room temperature. This has been achieved by partial replacement of Ag^+ by Rb^+. In $RbAg_4I_5$, the Ag^+ ions can also move freely between the anions, though the Rb^+ ions are fixed. There seem to be two reasons for this unusual behaviour of AgI in contrast, for example, to LiI which also has a small cation. Firstly, the structures are dominated by I^- packing because the cations are small. Secondly, Ag^+ forms partly covalent bonds to I. As a result, positions with low coordination numbers (4, 3, or 2) are stabilized. The Ag^+ ions can therefore move from one position to another with a smaller energy barrier than is possible for Li^+.

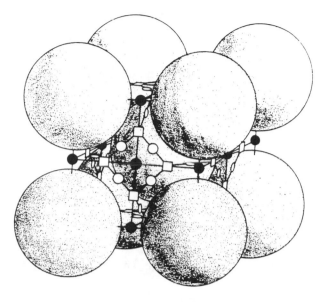

Fig. 10.8 — Structure of α-AgI. Reproduced with permission from Funke, K. (1976) 'AgI-type solid electrolytes', *Progr. Solid State Chem.* **11**, 345, p. 358.

10.6 FERROELECTRICS

The characteristic property of the superionic conductors is the ability of some of their ions to move through the crystal under the influence of an electric field. If this happens, but the ions move only short distances, then the compound may be a *ferroelectric*. This property was named by analogy with ferromagnetism (Box 5). It therefore describes a material in which groups of atoms have permanent electric dipoles and the individual dipoles are spontaneously aligned parallel to each other, like the individual magnetic moments in a ferromagnetic material.† In parallel to antiferromagnetism, *antiferroelectric* behaviour is also found, when the dipoles are aligned *anti*-parallel. In principle, above their Curie temperatures these materials become *para*-electric (with the dipoles randomly aligned), though in reality a structural change often takes place at this temperature and the permanent dipoles disappear. Ferroelectrics find crucial applications in electrical devices, especially in large capacitors where the energy of interaction with the dipoles contributes to the ability of the capacitor to store charge. They also show a closely related property, that of *piezo*-electricity, the development of electrical polarization when subjected to a mechanical stress (and the reverse of this). This is applied in microphones, loudspeakers, and stereo pick-ups.

The most important feature of ferroelectric compounds is that strong electric fields will reverse the direction of the dipoles. The dipoles are caused by the unsymmetrical arrangement of the atoms in the structure. Their reversal therefore means that the atoms have to move in the crystal, to take up the opposite

† The term *ferro*electric therefore has nothing to do with the material containing iron.

unsymmetrical arrangement. An essential structural feature of ferroelectrics is that the crystals do not contain centres of symmetry. This is found systematically for two classes of compounds: H-bonded salts and mixed metal oxides with distorted structures. Compounds containing zwitterions, e.g. $^+H_3N(R)CO_2^-$, may well pack in their crystals to give oriented dipoles, but they cannot reverse direction unless the whole ion rotates end-for-end. They are therefore not ferroelectrics.

10.6.1 Hydrogen-bonded ferroelectrics

The unsymmetrical nature of most H-bonds

$$O-H\cdots O$$

gives them an opportunity to switch to

$$O\cdots H-O$$

if the two O-atoms are chemically identical. This happens in KH_2PO_4 (Fig. 10.9).

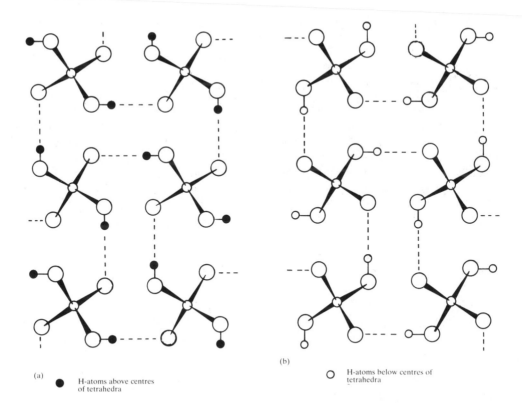

(a) ● H-atoms above centres of tetrahedra

(b)

○ H-atoms below centres of tetrahedra

Fig. 10.9 — Structure of KH_2PO_4, containing H_2PO_4 tetrahedra linked by H-bonds (a) polarized in one direction with all H-atoms above the centres of the tetrahedra, (b) polarized in the opposite direction with all H-atoms below the centres of the tetrahedra.

The structure contains $[H_2PO_4]^-$ tetrahedra, each forming four hydrogen bonds to other tetrahedra. At room temperature, these H-bonds are arranged randomly. On cooling to 123 K, a structural change occurs and the H-atoms become ordered with the short O–H bonds always on one side of the PO_4 tetrahedra associated with the upper oxygen atoms (Fig. 10.9a). When a strong electric field is applied, the protons move in concert to the other side of the O–H$\cdots$O bonds, reversing the polarity of the crystal (Fig. 10.9b).

10.6.2 Perovskite ferroelectrics

The distortions often found in compounds with the perovskite structure provide another mechanism for ferroelectricity. These distortions generally involve the BO_6 octahedra, rather than the AO_{12} units. Sometimes the octahedra twist in relation to one another; this happens in $SrTiO_3$ below 108 K (Fig. 10.10a), but the structure remains symmetrical, without each octahedron developing a dipole. This material is therefore not ferroelectric. Another common distortion involves the B-ions moving off-centre in the octahedra. In $BaTiO_3$ below 120°C, the Ti^{4+} ions are displaced from the octahedron centres by 12 pm. All the shifts are in the same direction, so this is a ferroelectric (Fig. 10.10b). In a strong electric field, the Ti^{4+} ions move from one side of the octahedron to the other, reversing the direction of polarization.

In contrast, in $PbZrO_3$, the Zr^{4+} ions move in opposite directions in each octahedron, so this is antiferroelectric (Fig. 10.10c). Interestingly, a *strong* electric field causes all the Zr^{4+} ions to move in the same direction, so that the material becomes ferroelectric. When the field is removed, the antiferroelectric arrangement reappears.

10.7 NON-LINEAR OPTICAL MATERIALS

Unusual optical properties in crystals depend on the overall character of the structure, rather than on the interactions between the individual atoms. They arise as special forms of the general interaction between light and matter. When visible or near visible radiation strikes a material, the oscillating electric field of the radiation interacts with the electrons in the material, causing them to oscillate, producing a rapidly varying dipole. The size of this dipole (P) is ideally given by the simple *linear* equation

$$P = \varepsilon_o \chi_1 E$$

where E is the electric field (depending on the intensity of the light), ε_o the permittivity of free space, and the parameter χ_1 is the 'linear electric susceptibility' of the material. This is directly related to its refractive index (n) by

$$n^2 = 1 + \chi_1.$$

In real materials, P may be rather more complicated:

$$P = P = \varepsilon_o(\chi_1 E + \chi_2 E^2 + \chi_3 E^3 + \ldots). \tag{10.2}$$

(a) Sr Ti O$_3$

(b) Ba Ti O$_3$

(c) Pb Zr O$_3$

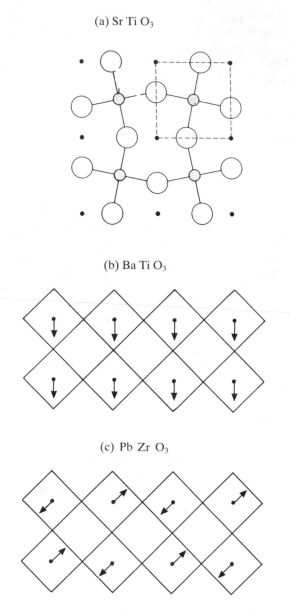

Fig. 10.10 — Distorted perovskites (a) twisted octahedra in SrTiO$_3$ (not ferroelectric), (b) small parallel diplacements in BTiO$_3$, a ferroelectric, (c) small antiparallel displacements in PbZrO$_3$, leading to the dipoles cancelling, i.e. anti-ferroelectric behaviour.

χ_2, χ_3, etc. are the second order, third order, etc. non-linear susceptibilities. Often χ_2 and χ_3 have values which are *precisely* zero (see below). If they are non-zero, then the material has non-linear optical properties. The effects obviously depend on the intensity of the radiation (E). Thus, they become important for intense sources, especially lasers. The effect of χ_2 can be seen if E is rewritten in the form of a wave (frequency ν)

$$E \sin(2\pi\nu t) \tag{10.3}$$

Substituting in (10.2) gives for the χ_2 term

$$\varepsilon_0 \chi_2 E^2 \sin^2(2\pi\nu t)$$

and replacing $\sin^2(\theta)$ by $1/2(1 - \cos^2(2\theta))$ gives

$$1/2\varepsilon_0\chi_2 E^2 - 1/2\varepsilon_0\chi_2 E^2 \cos^2(2\pi(2\nu)t)$$

This is a similar wave to the incoming one, but at *double the frequency*. Thus, if radiation of one wavelength, e.g. from a neodymium laser (1064 nm) falls on an optically non-linear crystal, radiation of half the wavelength is emitted, converting the original infrared to green (532 nm). This is of great practical importance in obtaining laser radiation at frequencies in the visible and u.v. spectrum for which no direct laser sources are available.

10.7.1   Materials with non-zero χ_2

The criterion for a material with non-linear optical properties is remarkably simple. It is that the crystals of the material do not contain a centre of symmetry.† To achieve this is less straightforward. The only way to be sure that a crystal lacks a centre is for it to contain an optically active compound, such as an l-amino-acid; if a centre were present, it would relate a molecule of the l-amino-acid to its d-form. For the most part, crystals of non-optically active materials crystallize with centres of symmetry. However, if the material is itself not very symmetrical, it will sometimes form a mixture of left- and right-handed crystals. The individual crystals are then potentially useful as a non-linear optical material.

A good example is $LiNbO_3$, in which the NbO_6 octahedra have three Nb–O distances of 188.9 pm and three of 211.2 pm (Megaw 1968).‡ This is one of the main non-linear optical materials in current use. A group of compounds of rather similar structure to this are the iodates, MIO_3 (e.g. $LiIO_3$). The 'octahedra' of oxygen atoms around each iodine are highly distorted (I–O distances about 180 and 280 pm; section 7.5). Their crystals often lack centres of symmetry and also have high χ_2 values. An important recent addition to these materials is $KTiO(PO_4)$, potassium titanyl phosphate.

As the previous discussion indicates, ionic compounds have no monopoly of non-linear optical behaviour. Indeed, it occurs in them only as the occasional result of structural distortions. However, they do have mechanical strength and thermal stability which are useful attributes for these materials. Recently, the search for such

---

† The same criterion applies to ferroelectric materials, but in addition they have to be able to reverse their handedness under the influence of an electric field.
‡ The structure of $LiNbO_3$ is a fairly large distortion of perovskite, and it is often described as related to ilmenite (Ti_2O_3).

compounds has turned to covalent compounds. These can offer an additional property, that of readily polarizable electrons, e.g. in π orbitals, which can give very high values of χ_2. In addition, by incorporating a chiral centre in the molecule, the growth of crystals without a centre of symmetry can be guaranteed; it has been also possible to overcome the problems caused by their low melting points and possible chemical reactivity (Pugh & Sherwood 1984).

10.8   MAGNETIC INTERACTIONS

The electrical interactions described in sections 10.5 and 10.6 were caused by the movement of individual ions. With magnetic interactions, no movement occurs, but the unpaired electrons on transition metal ions influence each other in remarkably complicated ways. These depend very strongly on the details of the structure and the distances between the ions. As a result, apparently identical compounds can be magnetically very different. The main types of behaviour are those already described (Box 5): antiferromagnetism and ferromagnetism, in which the magnetic moments (electron spins) of individual ions are aligned either anti-parallel (net magnetic moment zero) or parallel (non-zero net moment). Both systems become random above some critical temperature (the Neél or Curie temperature respectively).[†]

For technological applications, ferromagnetic materials are important, and much research is concerned with producing materials with high magnetic moments and Curie temperatures well above room temperature. A crucial parameter is the magnetic field needed to reverse this direction of magnetization. This should be as high as possible for a permanent magnet, but low for an electromagnet or transformer core; magnetic computer memory devices must have intermediate values, easily switched but not so low as to be affected by stray fields. These properties are influenced not only by the chemical composition, but also by the particle size and orientation, so preparation procedures need careful investigation and control.

10.8.1   Types of interaction

The simplest magnetic interaction to understand is that known as *superexchange*. It occurs when two ions with unpaired spins (e.g. Ni^{2+}) form a weak covalent interaction with an intermediate anion, such as O^{2-} (Fig. 10.11a). The interaction between the first Ni^{2+} and one electron on the O^{2-} causes these electrons to pair. Naturally, the two electrons on the O^{2-} are paired, so the interaction with the second Ni^{2+} leads to its electron pointing in the opposite direction to that on the first Ni^{2+}: *antiferromagnetic coupling*. This is the conventional but rather cumbersome description. The interaction can better be looked at from another viewpoint, as a further example of a three-centre/four-electron interaction (section 5.9.2) (Fig. 10.11b). One orbital on each Ni^{2+} overlaps with one on O^{2-}, to give three molecular orbitals, containing four electrons. It is obvious from this description that the two Ni^{2+} electrons should be paired, but it is less clear what happens to the interaction if the overlap is very small. In that case, the gain in energy from the formation of bonding molecular orbitals does not exceed the repulsion between the two electrons that

---

† The theory of magnetic interactions is complex and will be discussed only in outline. For further detail, see Guinier & Jullien (1989), Newnham (1975), West (1984).

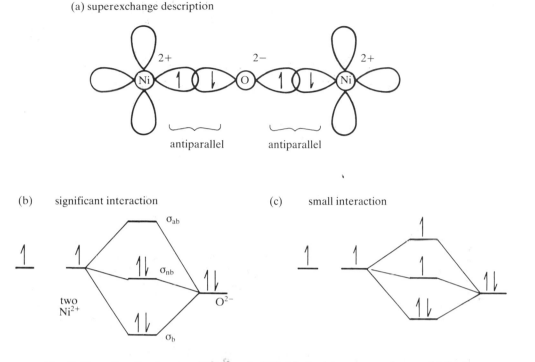

(a) superexchange description

antiparallel   antiparallel

(b)   significant interaction   (c)   small interaction

σ_{ab}

σ_{nb}

two
Ni^{2+} O^{2-}

σ_b

Fig. 10.11 — Coupling between Ni^{2+} ions in NiO (a) viewed as superexchange, (b) 3c–4e
bonding interaction, (c) 3c–4e interaction with very weak overlap.

occupy one orbital. The unpaired state is therefore preferred (Fig. 10.11c). Ferromagnetism might then result, but paramagnetism is more likely with the spins
pointing in random directions, because the interactions will be very small.

Experimentally, M–O–M interactions are generally strong, so most metal oxides
are antiferromagnetic with quite high Neél temperatures. For the simple MO oxides,
these temperatures increase in the order Mn>Fe>Co>Ni. The decrease in size of
the metal ions and their increasing nuclear charge leads to greater orbital overlap and
stronger interactions.

Ferromagnetic interactions are more complex. They occur in the transition
metals themselves that have the highest numbers of unpaired electrons (see section
6.3.2). They can also arise in complex oxides, including spinels. Here, the existence
of octahedral and tetrahedral sites produces metal ions with varying magnetic
properties. An interesting example is Fe_3O_4 (magnetite, the original lodestone).
This is an inverse spinel (section 10.2.1), with Fe^{2+} and Fe^{3+} ions in the octahedral
sites and Fe^{3+} in the tetrahedral sites. The octahedral Fe^{2+} and Fe^{3+} sites are not
distinct, and electrons move from d^6 Fe^{2+} to d^5 Fe^{3+}, exchanging their oxidation
states (Fig. 10.12). This can happen only if the electrons on each ion are parallel; if
not, one of the electrons would end up in an excited state and the process would have
a high activation energy.

Fe^{2+}/Fe^{3+} exchange in Fe_3O_4

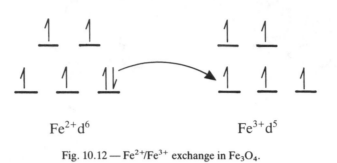

Fig. 10.12 — Fe^{2+}/Fe^{3+} exchange in Fe_3O_4.

Neutron diffraction (section 2.2.2) allows the alignment of electron spins to be studied in detail. This has shown that their arrangements are often very complex and difficult to explain. In particular, the spins on distant metal ions influence each other, leading to the repeating unit of the spins (the 'magnetic unit cell') being larger than the repeating unit of the atoms (the crystallographic unit cell). The three compounds MnS_2, $MnSe_2$, and $MnTe_2$ illustrate the subtle differences between similar compounds (Fig. 10.13). All have the FeS_2 (pyrites) structure, with X_2^{2-} units and M^{2+} ions in the unit cell, but the spins are arranged differently in the three crystals. For $MnTe_2$, the magnetic unit cell is the same as the crystallographic unit cell, but for MnS_2 the magnetic cell is doubled and for $MnSe_2$ trebled.

These are by no means the most complex examples known. The perovskite $BiFeO_3$ is antiferromagnetic, and the spins are arranged in a spiral (some pointing up and some down), which takes 62 000 pm for a complete rotation (150 times the unit cell length in the crystal) (Sosnowska *et al.* 1982). Such arrangements appear because the different interactions between the spins cannot all be satisfied; this leads to what is known as *frustration*. Typically, neighbouring spins tend to be parallel, with next-nearest neighbours antiparallel. Considering three adjacent atoms, the nearest neighbour interactions will set spins 1 and 2, and 2 and 3 parallel. However, spin 3 will also seek to be antiparallel to 1. It is in a *frustrated* state. The result will be that none of the spins are parallel, but they form a spiral. The angle between each pair of spins will minimize the energy for both types of interaction.

10.9 SUPERCONDUCTING METAL OXIDES

This section considers the most exciting structure/property relationship discovered in recent years, the superconductivity of complex metal oxides, whose critical temperatures (T_c; Box 12) are up to 100 K higher than for any other superconductors. Superconductivity was encountered briefly in considering metals, and will reappear in Chapter 11, but is examined in most detail here.

10.9.1 Principles of superconductivity
In normal metals, movement of electrons is impeded by their interactions with impurities and with lattice vibrations; these scatter the electrons, causing them to

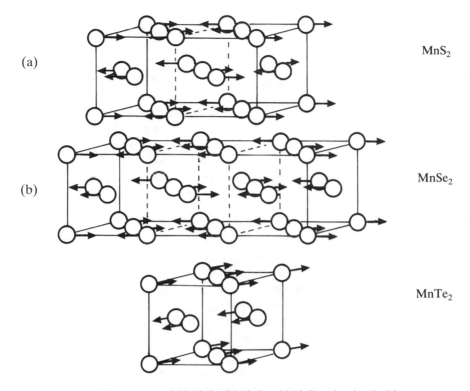

(a)   MnS$_2$

(b)   MnSe$_2$

                                                                       MnTe$_2$

Fig. 10.13 — Magnetic structures of (a) MnS$_2$, (b) MnSe$_2$, (c) MnTe$_2$, showing the Mn atoms and the directions of the magnetic spins. In each drawing, solid lines show the repeating unit (unit cell) of the magnetic structure, and the dotted lines mark the crystallographic non-magnetic unit cells. Source: data of Hastings, J. M., Elliott, N., Corliss, L. M., *Phys Rev*. **115**, 13 (1959).

lose energy. A perfectly pure and perfectly rigid metal single crystal would have zero resistance; of course, this cannot be achieved in practice, though very pure copper at 4 K has a resistance 10^{-5} times the normal value at room temperature. However, in superconducting materials the resistance is *exactly* zero. In some way the electrons have ceased to be scattered.

For superconducting metals, the BCS theory (Bardeen–Cooper–Schreiffer) explains how this arises. The theory shows that some of the conducting electrons are coupled in pairs ('Cooper' pairs) that are linked with a 'phonon', a quantum of the vibrational energy of the lattice. As a consequence, the possible energies of these pairs of electrons are restricted, and a substantial energy gap exists between this ground state and the lowest possible excited state, in which they behave as individual (normal) electrons. When this gap exceeds the available thermal energy, the electrons have insufficient energy to reach the excited state. Therefore they cannot be affected by the lattice vibrations and imperfections — they are superconducting.

As the temperature rises the gap decreases, vanishing at T_c when superconductivity ceases. The BCS theory allows T_c to be calculated, but only with detailed knowledge of the electronic structure and the lattice vibrations of the metal. It is difficult to obtain this information experimentally and impossible to predict it for an unknown material. The BCS theory cannot therefore identify possible new super-conductors. Correlations between T_c and the number of valence electrons (as in Fig. 6.15) are rather more useful, and have helped to identify some of the best superconducting metal alloys.

The difference in T_c between the highest values for metals (23 K for Nb_3Ge) and those for the metal oxides is so large that it is considered doubtful if the BCS theory can be applied to the oxides. It is presumed that electron pairing is still responsible for the superconductivity, but this may not involve interactions with lattice vib-rations. No accepted alternative explanation has been proposed, so the structures themselves provide the best evidence for the key features needed for superconductivity.

10.9.2 Structural types

The field of oxide superconductors has developed very rapidly in just a few years from late 1986, when the first Cu–O containing superconductor was discovered. During this period an enormous number of different oxide compositions have been examined, both to identify new superconducting phases and to achieve maximum T_c. From all these materials, four main families of superconducting oxides have been identified. They are listed in Table 10.2 with T_c values for typical compounds.

Table 10.2 — Superconducting metal oxides: formulae, structures and critical temperatures

	Typical Formula[a]	T_c/K	Structure type and comments
A	$Ba_{0.6}K_{0.4}BiO_3$ $[Ba_{1-x}K_xBiO_3]$	27	Perovskite
B	$La_{2-\delta}CuO_4$ $[La_{2-x}(Ba,Ca,Sr)_xCuO_4]$	40	K_2NiF_4. La<2 is necessary for superconductivity
C	$YBa_2Cu_3O_7$	90	Triple perovskite-related cell The '1:2:3 superconductor'
D	$Tl_2Ba_2CuO_6$	80	$n=1$
E	$Tl_2Ba_2CaCu_2O_8$	110	$n=2$ n layers of $[CuO_2]$ separated by $[ABO_2]$
F	$Tl_2Ba_2Ca_2Cu_3O_{10}$ $[A_2^{3+}B_2^{2+}Ca_{n-1}Cu_nO_{4+2n}]$	125	$n=3$ A=Tl,Bi; B=Sr,Ba

[a] Formulae are quoted for specific examples, but in all cases many similar compounds have also been examined.

10.9.2.1 $Ba_{1-x}K_xBiO_3$ (A in Table 10.2)

The first example in Table 10.2 is worth particular attention, even though its T_c is relatively modest (but still significantly higher than the maximum for any metal alloy). The proportion of potassium can vary in the range $0.25 < x < 0.4$, and it seems to have two functions. Firstly, it stabilizes an undistorted ABO_3 cubic perovskite structure (Fig. 10.4), with K and Ba randomly occupying the A sites (12-coordinate). $BaBiO_3$ itself and compounds with $x < 0.25$ have monoclinic distortions of the perovskite structure and are not superconducting. Secondly, substitution of K^+ for Ba^{2+} changes the average Bi oxidation state. Initially, this is 4+, presumably as a mixture of Bi^{3+} and Bi^{5+}, and addition of K^+ must increase the proportion of Bi^{5+}. The T_c is highest for $x = 0.25$ and drops as x increases. If alternative substituents can be found to stabilize the cubic phase with the mean Bi oxidation state nearer to +4, they may well give higher T_c values.

10.9.3   Cu–O superconductors
The three families of Cu–O superconductors are also structurally related to perovskite, though their key feature seems to be the precise copper ion environment.

10.9.3.1 $YBa_2Cu_3O_7$ (C in Table 10.2)
The most obvious relationship with the perovskites can be seen in the 1:2:3 superconductors† such as $YBa_2Cu_3O_7$. The unit cell of this oxide contains three perovskite-type cells stacked on top of each other, with Ba, Y, and Ba as their central A-atoms (Fig. 10.14a). An ideal triple perovskite would contain nine O-atoms, and

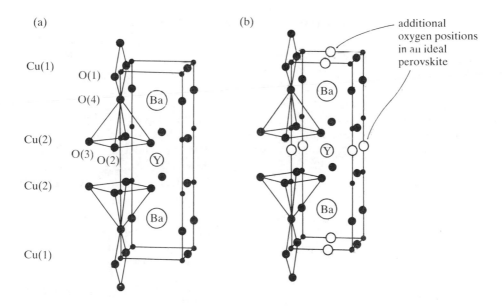

Fig. 10.14 — Structure of $YBa_2Cu_3O_7$ (a) actual structure, (b) showing positions of oxygen atoms missing in comparison to an ideal perovskite structure. Reproduced with permission from Edwards, P. P., Harrison, M. R., & Jones, R. (1987) Superconductivity returns to chemistry. *Chem. in Britain*, p. 963.

† So-called from the stoicheiometry of its metal ions. A variety of lanthanides can take the place of Y, and smaller amounts of other metals can be substituted for some of the Ba and Cu.

so this cell has two vacant oxygen positions. The oxygen content can easily be varied by heating either in oxygen or in a vacuum. As the number of oxygen atoms per cell increases from 6.5 to 7.0 (the maximum for which the structure is stable), T_c rises from 20 K to its maximum of 90 K. Compounds with 6.0–6.5 oxygen atoms per cell can also be made, but are semiconductors.

In comparison to an ideal perovskite (Fig. 10.14b), O-atoms are missing from two locations: a central belt of four oxygens around each Y, making it 8- rather than 12-coordinate, and two around each Ba, making them 10-coordinate. The effect on the three Cu ions in each cell is more complex. One Cu (called Cu(1)) is situated between two Ba layers where two unit cells are in contact. This Cu has four O-neighbours in a vertical plane, two of which are absent when the O-content is only 6 atoms per cell. The remaining pair of Cu atoms (Cu(2)) lie between the Ba and the Y layers. Each has four O-neighbours in a horizontal plane, and a more distant one above or below, giving it square-pyramidal geometry. The sixth oxygen position that would complete an octahedron if this was a regular perovskite is the one missing from the Y-coordination.

As well as alleving the geometry around Cu(1), the changing oxygen content must affect the copper oxidation state (assuming the oxygen remains as O^{2-}). With 6 O-atoms, all the copper is Cu^{2+}, but as the oxygen content increases, up to 1/3 of them become Cu^{3+}. It appears mainly to be Cu(1) which is oxidized, but Cu^{3+} is also likely to replace some Cu^{2+} on the Cu(2) sites; hopping of electrons from Cu^{2+} to Cu^{3+} is the probable cause of the metallic conductivity this compound shows above T_c. Bond valence calculations confirm that the oxidation state of Cu(2) is of crucial importance (Cava 1990). While the oxygen content changes from 6.5 to 7.0, the bond valence for Cu(2) increases rather irregularly. The variation in T_c precisely mirrors the variation in bond valence (Fig. 10.15).

10.9.3.2   Other Cu–O superconductors

The structures of the other families of Cu–O superconductors pinpoint the $\{CuO_2\}_n$ layers as being crucial for superconductivity (shown in Fig. 10.16). In $La_{2-\delta}CuO_4$ (B in Table 10.2), these layers alternate with LaO layers of NaCl structure. Thus, the formula can be written $\{LaO\}_2\{CuO_2\}$. As with the 1:2:3 superconductors, some Cu^{3+} is necessary, and this is achieved either by a deficiency of La^{3+}, or by replacing some La^{3+} by 2^+ ions (Ba, Ca, Sr).

In $Tl_2Ba_2CuO_6$ (D in Table 10.2) we find the same $\{CuO_2\}_n$ sandwich structure, but with a rather more complicated slice comprising $\{TlBaO_2\}_2$: a layer of BaO on each side of two layers of TlO, again with a NaCl arrangement; compounds with one TlO layer are also superconducting. Its formula can be written $\{TlBaO_2\}_2\{CuO_2\}$ (Fig. 10.17a).

Addition of Ca^{2+} to this structure allows the $\{CuO_2\}$ layers to stack together, interleaved with Ca^{2+}. The compound containing one such Cu–Ca–Cu double layer is $\{TlBaO_2\}_2\{Cu_2O_4Ca\}$ and adding a further Ca–Cu layer produces $\{TlBaO_2\}_2\{Cu_3O_6Ca_2\}$ (E and F in Table 10.2; Fig. 10.17b–c). The critical temperatures increase along the series, with (F) having the highest value of any fully characterized compound (125 K). The final member of the series has also been made, with an infinite set of $\{Cu_2O_4Ca\}$ layers and no intermediate NaCl portions. This is $Ca_{0.86}Sr_{0.14}CuO_2$ (the mixture of Ca^{2+} and Sr^{2+} being needed to stabilize the structure). Disappointingly, it is an insulator, but this is important in relation to the

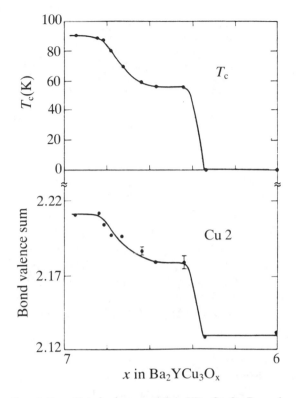

Fig. 10.15 — Correlation of bond valence and T_c in $YBa_2Cu_3O_x$. Reproduced with permission from Cava, R. J. (1990) Structural chemistry and the local charge picture of copper oxide superconductors. *Science* **247**, 656, copyright 1990 by the AAAS.

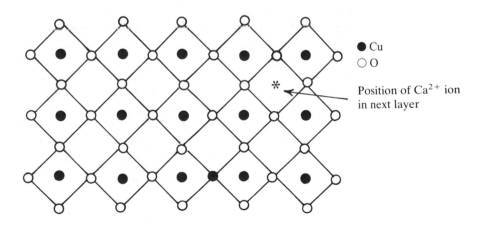

Fig. 10.16 — $\{CuO_2\}$ layer in Cu–oxide superconductors. In the $\{CuO_2Ca\}$ stacks, the Ca^{2+} ions are located between the empty squares in two layers (at the positions marked *). They therefore have cubic 8-coordination.

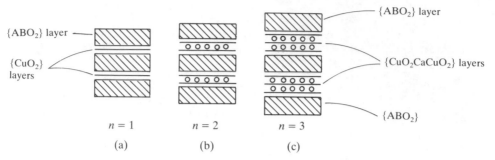

Fig. 10.17 — Schematic representation of $\{CuO_2\}$ and interleaving layers of NaCl structure (a) single $\{CuO_2\}$ layer (b) double layer of $\{CuO_2CaCuO_2\}$ (c) triple layer $\{ABO_2\}$. Reproduced with permission from Greaves, C. (1988) *Nature* **334** 193, copyright Macmillan Magazines Ltd.

properties of the other members of the series. Its insulating character is probably the result of its Cu atoms all being 2+ ions. Although compounds D–F also apparently contain only Cu^{2+}, evidence has been found that electrons can be transferred from the $\{CuO_2\}$ layers to the $\{ABO_2\}_2$ layers. The possibility of mixed Cu^+/Cu^{2+} ions has also been suggested. It is certainly significant that these compounds are superconducting only when A = Tl or Bi, and there may be a relationship with the superconductivity of $\{Ba,K\}BiO_3$ (compound A). The existence of this compound also suggests that other superconducting oxides not containing copper remain to be discovered.

The $\{CuO_2\}$ layer in these structures can be regarded as a fragment of a perovskite with some missing oxygen, particularly when it is interleaved with layers of Ca^{2+}. Indeed, if Fig. 10.14 is compared with Fig. 10.17b, we see that in $YBa_2Cu_3O_7$ the $\{CuO_2\}Y\{CuO_2\}$ layers are identical to the $\{CuO_2\}Ca\{CuO_2\}$ layers just described. The only major difference between compounds (C) and (E) lies in the non-$\{CuO_2\}$ layer; it comprises Ba_2CuO_3, i.e. two BaO sandwiching one CuO rather than two TlO.

10.9.4 Superconductivity and structure

This section has investigated the structural character of the superconducting oxides in more detail than for the compounds discussed earlier. It shows how the perovskite structure appears in different disguises. More important, because the mechanism of superconductivity in these oxides is still uncertain, we can only pick out the similarities and differences between the various examples as indications of what may be their key structural features. Even though our understanding of them is still imperfect, it is easy to see why they are the most important and exciting inorganic compounds discovered for many years — an importance which was recognized by the extraordinarily rapid award of the Nobel prize for this work, only two years after the original discovery.

FURTHER READING

General sources

Guinier, A. & Jullien, R. (1989) *The solid state*. Oxford University Press, Oxford.
Newnham, R. E. (1975) *Structure-property relationships*. Springer-Verlag, Berlin.
West, A. R. (1984) *Solid state chemistry and its applications*. Wiley, Chichester.

Structures

Brown, I. D. (1978) Bond valences — a simple structural model for inorganic chemistry. *Chem. Soc. Reviews* **8**, 359.

Burdett, J. K., Price, G. D., & Price, S. L. (1982) Role of crystal-field theory in determining the structures of spinels. *J. Amer. Chem. Soc.* **104** 92.

Muller, O. & Roy, R. (1974) *The major ternary structural families*, Springer Verlag, Berlin.

O'Keefe, M. (1989) The prediction and interpretation of bond lengths in crystals. *Str. and Bonding* **71**, 162.

Wells, A. F. (1984) *Structural inorganic chemistry*, Oxford University Press, Oxford.

Zirconia

Garvie, R. C., Hannink, R. H., & Pascoe, R. T. (1975) Ceramic steel. *Nature* **258**, 703.

Superionic conductors

Funke, K. (1976) AgI-type solid electrolytes. *Progr. Solid State Chem.* **11**, 345.

Vincent, C. A. (1984) *Modern batteries*. Edward Arnold, London.

West (1984), Chapter 13.

Ferroelectrics

West (1984), section 15.4.; Newnham (1975); Guinier & Jullien (1989), Chapter 2.

Non-linear optics

Thomas, P. (1990) Inorganic non-linear materials. *Physics World* March 1990, 33.

Pugh, D. & Sherwood, J. N. (1988) Organic crystals for non-linear optics. *Chemistry in Britain*, p.544.

Magnetic properties

West (1984), Chapter 16; Guinier & Jullien (1989), Chapter 4.

Superconductivity

Edwards, P. P., Harrison, M. R., & Jones, R. (1987) Superconductivity returns to chemistry. *Chem. in Britain.*

Cava, R. J. (1990) Structural chemistry and the local charge picture of copper oxide superconductors. *Science* **247**, 656.

Rao. C. N. R. & Raveau, B. (1989) Structural aspects of high-temperature cuprate superconductors. *Acc. Chem. Res.* **22**, 106.

Text references

Brown, I. D. & Altermatt, D. (1985) *Acta Cryst.* **B41** 244.

Megaw, H. D. (1968) *Acta. Cryst.* **A24**, 583.

Sosnowska, I. Peterlin-Neumaier, T., & Steichele, E. (1982) *J. Phys. C* **15**, 4835.

Wu, K. K. & Brown, I. D. (1977) *Materials Res. Bull.* **8**, 593.

11

Intermediate bonding

11.1   INTRODUCTION

Compounds in the centre of the bond triangle (Fig. 11.1) cannot be described in terms of simple bond types, but share the properties of several: apparently composed of ions but having ions with the same charge in contact; apparently covalent but with conductivity between molecules or throughout the solid; apparently metallic but with some of the metal 'atoms' composites of cations and anions.

One of the most important properties many of these compounds share is the presence of electrons delocalized through the solid, leading to electronic conductivity (in contrast to the ionic conductivity described in section 10.5). This has major technical applications, for example in semiconductors. Other novel properties are the subject of intense current investigation, such as electron transfer between covalently bound organic ions. In a general view, the most important effect of changing bond character is on structure.

11.2   HOMOPOLAR COMPOUNDS WITH NEARLY COVALENT BONDING: SEMICONDUCTORS

The simplest compounds with intermediate bonding are those that lie near the diagonal of the bond triangle. These are composed of one element (e.g. Si, Ge) of fairly low electronegativity, or of a combination of such elements. As the bond triangle shows, they fall between the areas of metallic and covalent bonding, though at first sight, they resemble covalent compounds most closely. In particular they lack the metallic conductivity associated with delocalized electrons.

In their structures, Si and Ge are identical to diamond (Fig. 11.2), with each atom covalently bonded to four others in tetrahedral geometry. The atomic orbitals of one atom can be described as sp^3 hybrids, forming four filled bonding and four empty antibonding molecular orbitals. However, as with CH_4 (section 5.4), this is an oversimplification, as the individual orbitals interact with each other. The structure

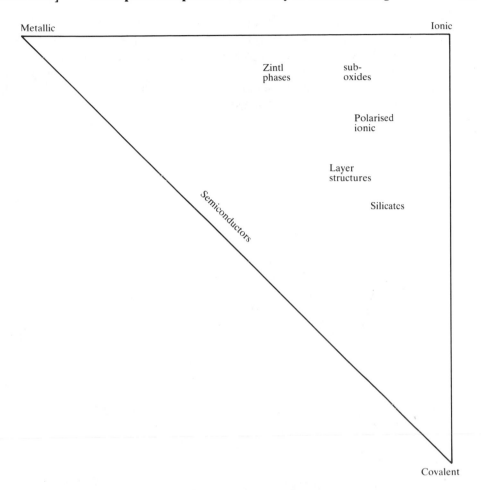

Fig. 11.1 — The bond triangle: compounds with intermediate bonding.

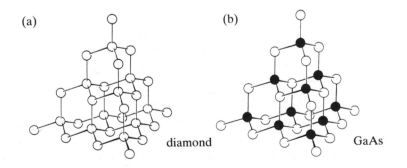

Fig. 11.2 — (a) The diamond structure, (b) structure of a III/V compound, GaAs.

as a whole therefore contains a *band* of bonding orbitals and a band of antibonding orbitals. As the $4n$ individual bonding orbitals of n atoms each contain two electrons, the bonding band (still containing $4n$ orbitals) is filled by the $8n$ electrons. The change on moving from C to Si to Ge is that the electrons are less strongly held. The molecular orbitals extend further from the individual atoms and overlap more with their neighbours. As a result, the bonding and antibonding bands become wider, and especially the gap (the *band gap*) between them becomes smaller (Fig. 11.3).

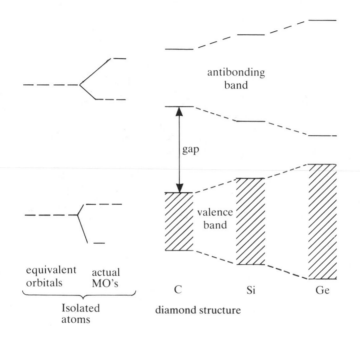

Fig. 11.3 — Band structure in C, Si, and Ge.

The experimental band gaps can be measured by absorption spectroscopy and are:

Atom	C	Si	Ge	Sn[†]
Band gap (eV)	5.5	1.1	0.7	0.1

† Sn in its unstable grey tin form (diamond structure)

The effect of the decreasing gap on the physical properties is dramatic. Diamond is an excellent electrical insulator, as are Si and Ge at low temperatures. However, in the latter at room temperature, some of the electrons are thermally excited and have enough energy to move from the bonding to the antibonding band. Fig. 11.4 shows this effect (plotting the number of orbitals ($N(E)$) against energy, as for metals in section 3.3).

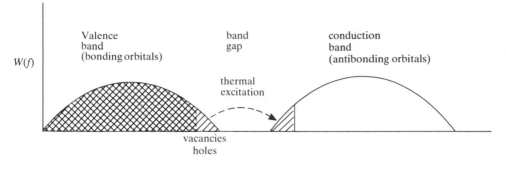

Fig. 11.4 — Bands in Si.

The few electrons in the antibonding orbitals are free to move in an electric field and therefore conduct electricity; this band is therefore called the *conduction band*. The filled band (the bonding orbitals) is known as the *valence band*. Following thermal excitation it contains some vacancies or '*holes*' in its highest energy levels. Holes behave in almost the same way as electrons in conducting electricity. An electron can be promoted to occupy one of the holes and so become mobile in an electric field. However, when it moves in one direction, filling a hole, the orbital it vacates becomes the hole, which therefore moves in the other direction. As a result, conduction by holes can be treated as conduction by positively charged electrons.

11.2.1 Intrinsic semiconductors
A material such as silicon which has a low conductivity that increases as the temperature rises, is called a *semiconductor*.† When the current is carried by equal numbers of excited electrons and holes, it is an 'intrinsic semiconductor', as are pure Si and Ge. Their conductivity (σ) depends on the number of electrons promoted and therefore the band gap (E_g) and the temperature (T), and also on the mobility of the electrons (μ) (11.1)

$$\sigma \propto \mu.\exp(-E_g/kT). \tag{11.1}$$

Only if the band gap is comparable to $1/kT$ (about $0.025\,\text{eV}$; $2.4\,\text{kJ mol}^{-1}$ at room temperature) will large numbers of electrons be promoted, giving a relatively high conductivity. Formula (11.1) is not very easy to examine experimentally. The mobility is affected by interaction between the conduction electrons and the vibrations of the lattice (as for metals) and is therefore also a function of temperature (an effect ignored in (11.1)). In addition, the electrons also interact with defects in

† In contrast, in metals the conductivity decreases as the temperature increases, because the constant number of conduction electrons undergo increasing interactions with the vibrating atoms in the crystal. These interactions are also found in semiconductors but have a smaller effect than the increasing number of charge carriers.

the lattice and with impurity atoms. The mobility therefore depends on the purity and perfection of the crystal. As will be seen in the next section, impurities can also have a great effect on the numbers of electrons and holes. Reproducible measurements of the conductivity of intrinsic semiconductors therefore place great demands on the quality of the material.

11.2.2   Doped semiconductors

The conductivity of a semiconductor can be greatly altered by the presence of atoms of other elements. Taking Si as an example, small amounts of either P or Al can be included in the crystal in solid solution (section 6.4.1), replacing some Si atoms without disturbing the structure. Each P atom has five valence electrons, but the valence band of the Si crystal is already filled by four electrons from each atom. The extra electrons can go only into the conduction band. Al atoms have the opposite effect, producing additional holes. In both cases the conductivity increases.

Semiconductors with excess electrons (*negative* charge carriers), e.g. Si doped with P, are known as n-type; those with excess holes (*positive* charge carriers) are p-type. As well as their increased conductivity, these materials acquire important electronic properties. This behaviour will not be considered in detail, but one example is given as an illustration of the applications of semiconductors. A junction between an n- and a p-type semiconductor will allow current to pass in one direction and not the other. It can be used for *rectification* of alternating current. Fig. 11.5a compares the energy levels of the two types of semiconductors. Their Fermi levels (the highest electron energy level at absolute zero) are different. As a result, if the two semiconductors are brought into contact, they are not in equilibrium. Electrons move from n- to p-, giving the n-type a positive charge, depressing its energy levels. The extra electrons raise the energy levels of the p-type material, so that the Fermi levels become equal. In the contact area, both holes and electrons are depleted, forming an insulating barrier. An external electric field is now applied either in one direction or the other. In one direction, it increases the energy difference between the two ends, enlarging the insulating barrier (Fig. 11.5c). In the other direction, the barrier is reduced or removed. Electrons from one side move to the other, combining with holes from the other side. Thus current flows moving only when the field is in this direction.

Doped semiconductors have many other important properties. They can emit light when electrons and holes recombine (semiconductor lasers), or convert light into electric current (solar cells). Their conductivity may also be increased by light (photoconductivity). This is applied in the Xerox process, in which a film of amorphous selenium is electrostatically charged. Light from the bright parts of the image discharges the corresponding portions of the selenium film. Powdered ink then sticks to the remaining parts of the film, is transferred to paper and fixed by heating. Photoconductivity is also applied in the detection of infrared radiation, in night-vision systems, and other applications.

These and other semiconductor properties all derive from the same general band structure, though the specific properties depend on the detailed nature of the interactions between energy levels, holes or conduction electrons, and external electrical fields. More details can be found in the sources listed under *Further reading*.

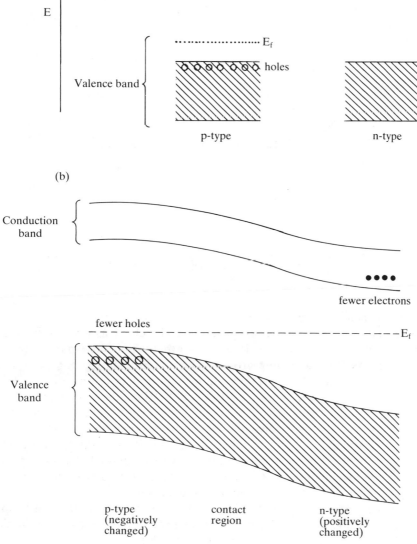

(a)

E

Conduction band

electrons

E_f

Valence band

E_f

holes

p-type

n-type

(b)

Conduction band

fewer electrons

fewer holes

E_f

Valence band

p-type
(negatively
changed)

contact
region

n-type
(positively
changed)

Fig. 11.5 — Energy levels of semiconductors (E_f=Fermi level) (a) isolated, (b) n- and p-type semiconductors in contact.

(c)   (d)

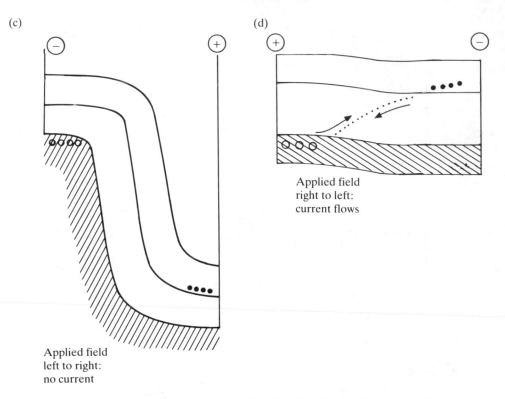

Applied field
right to left:
current flows

Applied field
left to right:
no current

Fig. 11.5 — Energy levels of semiconductors (E_f=Fermi level). (c) with applied field left to
right: no current (d) with applied field right to left: current flows.

11.2.3 Mixed element semiconductors

Silicon and germanium are the principal *elements* which behave as semiconductors,
but many compounds behave similarly. In particular these elements can be replaced
by pairs of elements from adjoining groups. These compounds generally retain
tetrahedral structures (Fig. 11.2b); as they still have four electrons per atom, their
valence bands are full. What are known as III/V and II/VI compounds include AlP,
GaAs, InSb, ZnS, and CdTe.† It is also possible to substitute these further, for
example giving $AgInTe_2$ as a derivative of CdTe. Other semiconductors achieve full
valence shells by forming bonds between two anions (or occasionally two cations).
Thus CdSb has seven electrons per formula unit in its valence band, but it contains an
Sb-Sb bond. This allows all three Sb–Cd bonding molecular orbitals to contain a pair
of electrons, and the valence band formed from them is full.

The principal effect of changing the elemental composition is on the band gap.
Increasing the electronegativity difference between the atoms increases the ionic
character of the bonding and gives a larger band gap. Thus

† These have yet to be recognized as 13/15 and 14/16 compounds to conform with the recent relabelling of
the groups!

$$\text{Ge (gap } 0.7\,\text{eV}) < \text{GaAs } (1.5\,\text{eV}) < \text{ZnSe } (2.8\,\text{eV}).$$

It is therefore possible to tailor the semiconductor properties for particular applications.

11.3 POLARISED IONIC BONDING
The logical next step from II/V and II/VI semiconductors is to I/VII compounds, such as NaCl. This was described in Chapter 4 as composed of Na^+ and Cl^- ions held together only by electrostatic forces, but the extrapolation from II/VI compounds suggests that NaCl should be a covalent solid with a rather large band gap!

Fortunately, we can reconcile these descriptions. Fig. 11.6 shows the effect of

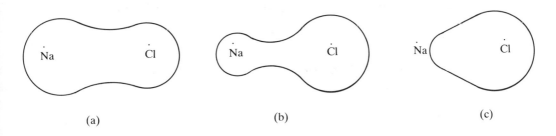

(a)   (b)   (c)

Fig. 11.6 — Change in the bonding molecular orbital in a NaCl bond as ionic character increases. For simplicity, these diagrams show an isolated pair of atoms, but the same changes occur in an extended structure. (a) No ionic character, (b) intermediate, (c) almost completely ionic.

increasing the ionic character in an NaCl atom pair, starting with a pure covalent interaction between the atoms. In this system, the bonding molecular orbital contains two electrons, the single Na 3s electron and the bonding Cl outer electrons. If the bond has no ionic character, the bonding molecular orbital is symmetrical in relation to the two atoms (Fig. 11.6a). As the bond becomes more ionic, so this orbital is concentrated on the Cl atom (Fig. 11.6b), until in a perfectly ionic bond, the bonding orbital is simply the Cl atomic orbital (Fig. 11.6c), which contains both electrons; the atom pair is then made up of Na^+ and Cl^- ions.

These changes can be used to define the *fractional ionic character* or *ionicity* (f_i) of a compound (Phillips 1970). If the bonding orbital is written as

$$\psi_b = a\psi_1 + b\psi_2 \tag{11.2}$$

then

$$f_i \text{ can be defined as } (a^2 - b^2)/(a^2 + b^2) \tag{11.3}$$

$(a^2 + b^2 = 1$ if the orbital coefficients are normalized; Box 7). The values of a and b are related to the energy difference between the two atomic orbitals. Taking this difference as E_i, and the energy splitting between the bonding and antibonding molecular orbitals as ΔE (Fig. 11.7), it can be shown that the ionicity is also given by

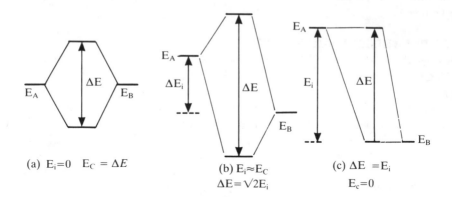

(a) $E_i=0$ $E_c = \Delta E$

(b) $E_i \approx E_C$
$\Delta E = \sqrt{2}E_i$

(c) $\Delta E = E_i$
$E_c = 0$

Fig. 11.7 — Energy levels of molecular orbitals in Fig. 11.6.

$$f_i = E_i / \Delta E \qquad (11.4)$$

ΔE can be written as

$$\Delta E = \sqrt{(E_i^2 + E_c^2)} \qquad (11.5)$$

in terms of two components, E_i and E_c, labelled as the ionic and covalent contributions to the orbital splitting (and hence the bonding energy). E_i has already been defined, and E_c is calculable from (11.5). Fig. 11.6 shows the orbital energies for the extreme cases of $f_i=0.0$ and 1.0 and for an intermediate situation.

This definition can be extended from diatomic molecules to extended solids with valence and conduction bands, by taking ΔE as the *average* energy difference between the two bands (i.e. between the middle of each band). This can be calculated from E_c, itself obtained by extrapolation from homopolar solids such as C, Si, Ge. ΔE can also be measured directly from the optical spectra of the compounds. Values of ionicity for some AB compounds show how it increases with the distance of the component elements from Group 14.

Compound	Ge	GaAs	ZnSe	CuBr	AgI	AgBr	LiF	NaCl
f_i	0	0.310	0.623	0.735	0.770	0.850	0.915	0.935

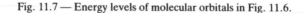

Diamond/ZnS structure   NaCl structure

Clearly, it is still reasonable to describe NaCl as ionic. However, this more detailed analysis shows that the identification of ionic rather than covalent bonding is not a matter of 'yes' or 'no', but of placing a particular compound on a continuous scale.

Variation in ionicity is particularly significant in relation to structure, especially the choice between an 'ionic' structure, such as the NaCl structure, or a more covalent one, like ZnS (cf. section 4.3.2). The most dramatic illustration of the structural significance of ionicity comes from a 'Phillips-van Vlechten plot' of E_i against E_c. Fig. 11.8 shows this plot for 68 AB compounds. It demonstrates that the

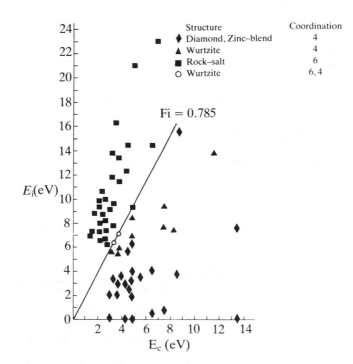

Fig. 11.8 — Phillips–van Vlechten plot for 68 AB compounds having either NaCl or ZnS (zincblende and wurtzite) structures. At $E_i=0$, the four group 14 elements are also included. Reproduced with permission from Phillips, J. C. (1970) *Rev. Modern Phys.* **42** 317.

change from the [6:6] NaCl structure to the [4:4] wurtzite/zincblende structures occurs exactly at $f_i=0.785$. The two compounds with precisely this value of f_i, MgS and MgSe, show both structures.

Thus, this structural change is controlled by the amount of ionic character in the bonding. As this decreases it becomes more favourable to form four fairly covalent bonds rather than six essentially ionic bonds.

The correlation of structure with other properties has also been investigated. A particularly effective separation in AB compounds is achieved in the 'Mooser-Pearson' plot, which uses the electronegativity difference and the average principal quantum number ($\bar{n}$ e.g. 2 for LiF, 4.5 for RbI) (Fig. 11.9). This plot shows that the CsCl/NaCl transition (section 4.3) can be seen as a consequence of changing chemical properties, rather than as a purely geometrical effect of changes in size. However, there is no obvious chemical rationalization for the appearance of the CsCl structure at the top of the Mooser–Pearson plot, like the increasing covalency which underlies the NaCl/ZnS transition. It is also worth remembering that size increases with principal quantum number, so the geometrical and chemical criteria are not entirely independent.

Such plots have now been extended to examine 20 different structural types in AB compounds (Villars 1983). Three parameters had to be used to separate these successfully, the electronegativity, the radius, and the average number of valence electrons. The separation was remarkably successful, with only 22 violations among 1000 compounds examined. The success of this type of correlation does not invalidate the classification by radius alone used for AB_2X_4 compounds (Fig. 10.1); most of these compounds contain oxide anions combined with metal cations of fairly constant electronegativity. With little variation in these factors, the size of the ions controls the structure.

These structural separations are important because they help to identify the factors on which crystal structures depend, but it is important to recognize their empirical nature. They do not provide chemical explanations of the distinctions they reveal.

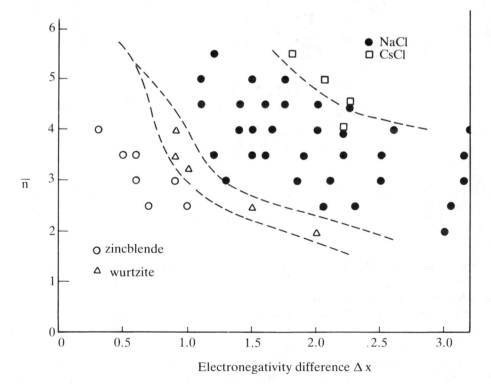

Fig. 11.9 — Mooser–Pearson plot for AB compounds containing a main group cation. Source: Mooser, E. & Pearson, W. B. (1959) *Acta Cryst.* **12** 1015.

11.4   LAYER STRUCTURES

The structural changes produced by the changing chemical character of the elements involved are not restricted to AB compounds, though these have been examined in most detail. In AB_2 and AB_3 compounds, the larger B:A atomic ratios, allow the adoption of structures inaccessible to AB compounds: layer and chain structures. The chlorides of Zn, Cd, Hg show clearly the effect of changing electronegativity. $ZnCl_2$ has tetrahedrally coordinated Zn in a [4:2] 3-dimensional structure related to the [6:3] coordinated TiO_2 structure in the same way as [4:4] ZnS is to [6:6] NaCl. $CdCl_2$ has a layer structure (next section), while $HgCl_2$ contains isolated linear molecules (Hg–Cl 225 pm), though four further long Hg–Cl interactions (334–363 pm) complete an octahedron; these should probably be regarded as weak secondary bonds.

11.4.1   The $CdCl_2$ and $CrCl_3$ layer structures

$CdCl_2$ has [6:3] coordination as does TiO_2. However, their resemblance ends there. In contrast to the latter's 3-dimensional network, the Cd atoms in $CdCl_2$ form a layer sandwiched between two layers of Cl atoms (Fig. 11.10). Next to the Cl-layer, we might expect more Cd. What we find is the Cl-layer of another Cl–Cd–Cl sandwich, with only van der Waals contacts between the Cl atoms of the different layers (368 pm). The structure consists of a stack of these sandwiches. This immediately shows that it cannot be significantly ionic. If the Cl-atoms carried substantial negative charges, the adjacent negatively charged layers would not be stable.

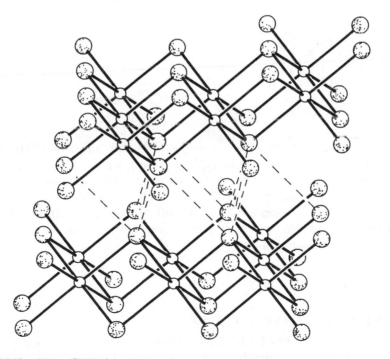

Fig. 11.10 — Linked $[CdCl_6]$ octahedra in layers of the $CdCl_2$ structure. The dashed lines mark some of the contacts between layers.

From a geometrical viewpoint, $CdCl_2$ an be regarded as containing cubic close-packed Cl atoms, with Cd atoms occupying half the octahedral holes. The unoccupied holes are those between the adjacent layers of Cl atoms. CdI_2 has an almost identical structure, except that the close-packed I atoms are stacked in the hexagonal (ABAB) arrangement, rather than the cubic (ABCABC) packing. These two structures are adopted by almost all the dihalides of the transition metals, apart from the fluorides (which have the rutile structure). Metal disulphides and selenides, e.g. TaS_2, also often show the $CdCl_2$ or CdI_2 structures.

Very similar structures are found for MX_3 compounds. $CrCl_3$ (Fig. 11.11) exists with two alternative stackings of Cl–Cr–Cl sandwiches, corresponding to the arrangement of the layers in $CdCl_2$ and CdI_2. The only substantial difference between $CrCl_3$ and $CdCl_2$ comes from the coordination ratios [6:2] instead of [6:3], leading to angular

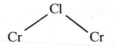

coordination instead of pyramidal

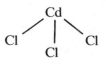

groups.

Metal trihalides can adopt linear as well as layer structures, as in ZrI_3. This has a very simple arrangement: octahedral $[ZrI_6]$ groups, with each I bridging between two adjacent Zr atoms (Fig. 11.12). All the transition metal trihalides adopt linear or layer structures (again excepting the fluorides). These structures seem to differ little in energy so that $ZrCl_3$, for example, shows both the $CrCl_3$ and ZrI_3 structures. Both these structures can also be described as having close-packed halogen atoms with metal atoms in octahedral holes.

11.5 INTERCALATION COMPOUNDS

The structures just described with van der Waals contacts between layers of halogen or sulphur atoms have a remarkable property: the ability to insert atoms, molecules, or ions between the layers. This is known as *intercalation*. Among metal compounds

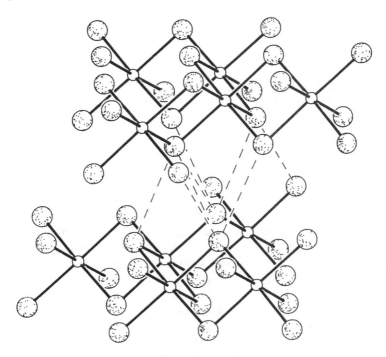

Fig. 11.11 — Linked [CrCl$_6$] octahedra in layers of the CrCl$_3$ structure. The dashed lines mark some of the contacts between layers. In comparison to Fig. 11.10, note how some of the Cl$_6$ octahedra contain no metal atoms.

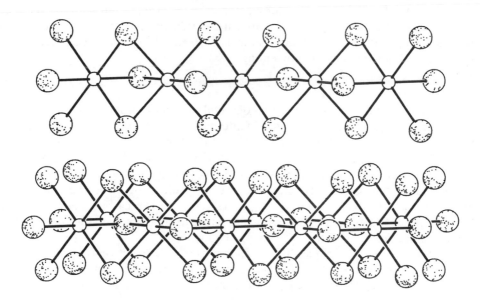

Fig. 11.12 — Linked [ZrI$_6$] octahedra in ZrI$_3$ chains.

it has particularly been studied for metal sulphides such as NbS_2, but it is also found for graphite and some silicates which have layered structures (the clays and micas, section 11.7.2).

11.5.1 Intercalated metal sulphides

A remarkable variety of species can be intercalated in transition metal sulphides and selenides. They range from alkali metal ions (isolated or hydrated) to NH_3 and organic amines. Generally, formation of the intercalation compounds involves transfer of electrons to the sulphide layer, so that they can be written as $2K^+[3NbS_2]^{2-}$ or $[EtNH_3]^+[Ta_2S_4]^-$. The structural effect of intercalation is that the MS_2 layers move apart and the intercalating atoms take up positions between them. Because the stability of a layer is little affected by the addition of electrons, the intercalated compounds are often non-stoicheiometric. Thus Li_xTiS_2 is found for all x between 0.0 and 1.0. Organic amines are aligned in pairs between the layers, so that the positively charged NH_3^+ ends are closest to the MS_2 layers (Fig. 11.13a). Other intercalates show a systematic variation known as 'staging'. The inserted atoms may occupy every gap or for lower metal: intercalate ratios every second, third, fourth, ... gap (Fig. 11.13b).

These compounds are of technical significance. The alkali metal ions are mobile within the gaps, and the materials therefore show ionic conductivity. However, the extra electrons in the MS_2 layers enter the conduction band, so that they are also electronic conductors.† They are therefore not useful as solid electrolytes (cf. section 10.5), but can function as cathodes in high power cells. Li_xTiS_2 performs very well in this role, reversibly taking up and losing Li^+.

Intercalation is also of interest in relation to superconductivity. TaS_2 itself is a metallic conductor that becomes superconducting below 0.8 K, but in its intercalate with $[K(H_2O)_x]^+$, T_c increases to 5.3 K. Remarkably, compounds with intercalated long-chain amines, e.g. $C_{18}H_{37}NH_2$, are superconducting, even though their inter-layer spacing is 5000 pm (ten times that in TaS_2 itself). This superconductivity must be a purely 2-dimensional phenomenon.

11.5.2 Graphite intercalation

Graphite intercalation is similar to that of metal sulphides, though it was discovered much earlier (C_8K in 1926). The range of intercalates is even wider: the alkali metals, transition metals, metal halides (e.g. $AlCl_3$, $NiCl_2$), non-metal halides, oxides and oxyacids (e.g. SO_3, HNO_3), Cl_2 and Br_2 (not F_2 or I_2 which are respectively too reactive and too unreactive). Their structures are similar to those just described, and sequences of compounds are often formed with different layer sequences. For example, C_8K, $C_{24}K$, $C_{36}K$, $C_{48}K$, $C_{60}K$ have one to five carbon layers between layers of potassium ions. The potassium atoms are located between the centres of C_6 hexagons in adjacent graphite layers (Fig. 11.14).

As with the metal sulphides, the alkali metals transfer electrons to graphite, e.g. $[K]^+[C_8]^-$. However, the electronegative intercalates remove electrons, giving such compounds as $[C_8]^+[Br]^-$ or $[C_{12}] \rightarrow SbF_5$, either completely ionized or with charge-

---

† The parent sulphides vary between insulating, semiconducting, and metallic behaviour, depending on their precise band structure.

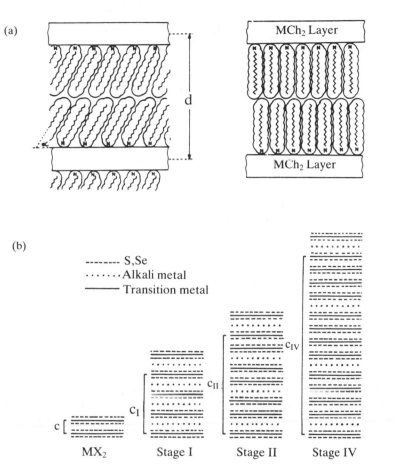

(a)

MCh$_2$ Layer

d

MCh$_2$ Layer

(b)

-------- S,Se
· · · · · · Alkali metal
———— Transition metal

c_{IV}

c_{II}

c_I

c

MX$_2$ Stage I   Stage II   Stage IV

Fig. 11.13   Intercalation structures, (a) amine intercalates (b) alkali metal intercalates, both in transition metal sulphides. Sources (a) Subba Rao, G. V. & Shafer, M. W. in Levy, F. (ed.) (1979) *Intercalated layer materials*. Reidel, Dordrecht, p. 138. (b) Rouxel, J. *ibid*, p. 210.1.

transfer interactions. Both electron transfer processes increase the number of charge carriers (either electrons or holes) in the graphite layers, leading to enhanced conductivity in the plane. Graphite itself is a moderately good 2-dimensional conductor (σ 1% of that for Cu), but intercalation increases this substantially; for the electron-acceptor intercalates, the conductivity can even exceed that of metallic copper.

11.6 LAYERED COVALENT COMPOUNDS: MOLECULAR METALS

The combination of a layered structure and the ability to gain or lose electrons is not restricted to graphite and metal sulphides. A number of planar unsaturated covalent

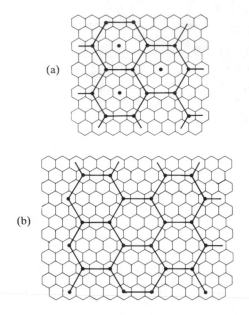

(a)

(b)

Fig. 11.14 — Two arrangements for metal/graphite intercalation. (a) for C_8M, (b) for $C_{12}M$.

molecules containing S, Se, or Te have been found to have intriguing properties which have led to their description as *molecular metals*. Of the examples shown (I–IV), the first to be studied was TTF (I). This can act either as an electron donor or an acceptor. It forms a 1:1 complex with TCNQ (V), an electron acceptor. Partial

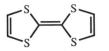

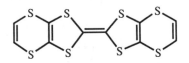

(I) TTF — tetrathiafulvalene.   (II) BEDT-TTF — bis(ethylenedithio)tetrathia-fulvalene.

(III) TMTSF — tetramethyl-tetraseleno-fulvalene.   (IV) Ni(dmit)₂ — nickel-bis(isotrithione-dithionate).

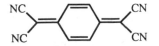

(V) TCNQ — tetracyano-quinodimethane.

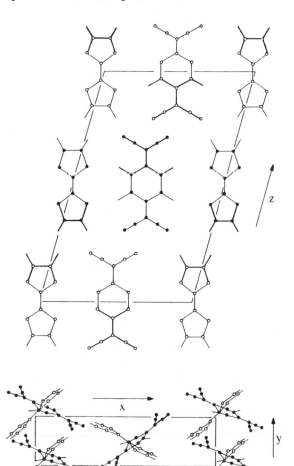

Fig. 11.15 — Crystal structure of the TTF:TCNQ complex. Reproduced with permission from
Jerome, D. & Schultz, H. J. (1982) *Adv. Phys.* **31** 299.

electron transfer takes place from TTF to TCNQ, and the complex crystallizes with
the TTF and TCNQ in columns, with all molecules parallel (Fig. 11.15). This packing
is characteristic of these species, and results from weak secondary bonding from the
sulphur (or selenium) atoms. The most important effect of this alignment is that the
highest occupied orbitals (the diffuse non-bonding lone pair orbitals on sulphur)
overlap between molecules, and are broadened into a band. Because of the transfer
of electrons to TCNQ, this band is part-filled. TTF:TCNQ therefore has a high
metallic conductivity ($\sigma = 10^4 \, \text{ohm}^{-1} \, \text{cm}^{-1}$) in one direction. As the temperature is
reduced, the conductivity shows the increase that is characteristic of a metal (Fig.
11.16). However, it reaches a maximum at 80 K, decreasing rapidly below that. This
is caused by a distortion of the structure. The TTF groups move alternately slightly
nearer and slightly further away from each other. As a result, what was one half-filled

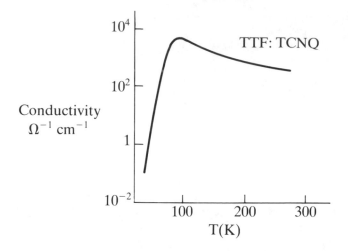

Fig. 11.16 — Temperature dependence of the conductivity of the TTF:TCNQ complex.
Reproduced with permission from Torrance, J. B. (1978) *Acc. Chem. Res.* **12** 79, copyright
American Chemical Society.

band becomes two bands, corresponding to the alternately better and worse overlap
of the atomic orbitals; Fig. 11.17 shows this schematically. At 80 K, the conduction
band is still thermally populated, and the conductivity remains high, but it rapidly
decreases as the temperature falls.

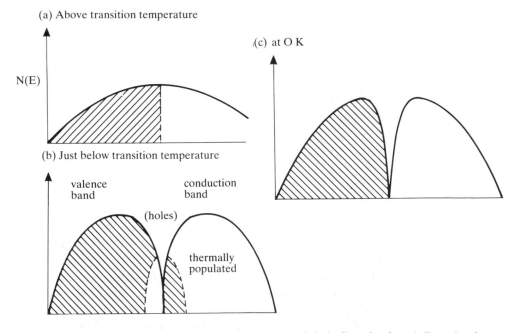

Fig. 11.17 — Metal-insulator transition caused by a Peierls distortion in a 1-dimensional
conductor: schematic view of the change in band structure (a) above transition temperature, (b)
just below transition temperature, (c) at 0K.

A distortion of this type is known as a *Peierls distortion*, and it can be shown that any *one-dimensional* metallic conductor *will* undergo such a distortion, causing it to become insulating when the temperature is reduced. Such distortions do not occur with 2-dimensional or 3-dimensional conductors.

Modification of the TTF structure and substitution of Se for S gives TMTSF (III), from which conducting salts such as $(TMTSF)_2PF_6$ can be produced (Fig. 11.18). In this salt, each TMTSF has lost 0.5 electron from its highest occupied molecular

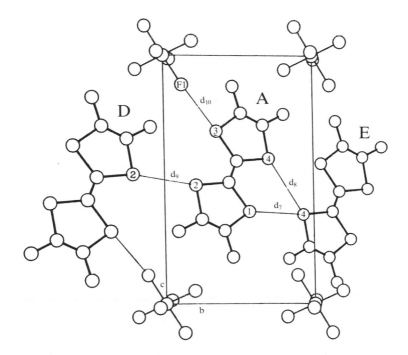

Fig. 11.18 — Structure of $(TMTSF)_2PF_6$. The distances d_7–d_{10} are the short interstack Se . . . Se interactions. Reproduced with permission from Williams, J. M. *et al.* (1983) *J. Amer. Chem. Soc.* **105** 643, copyright American Chemical Society.

orbital. The change in structure has a considerable effect on the metal-insulator transition which is now at 19 K. Furthermore, if pressure is applied (6.5 kbar), the transition is suppressed completely, and at 0.9 K the salt becomes superconducting. Like TTF:TCNQ, this salt contains stacked cations with overlapping orbitals (Se . . . Se distance around 370 pm). The key difference is that in $(TMTSF)_2PF_6$, the cations also form weak Se . . . Se contacts in the molecular plane (Se . . . Se around 390 pm). Under pressure, these contacts become strong enough for the material to be a 2-dimensional conductor, suppressing the Peierls distortion.

The exciting discovery of this new class of superconductors (in 1980) was followed by the synthesis of many related compounds. Changing the anions in $(TMTSF)_2PF_6$ changes the packing of the cations slightly, and affects the intermolecular interactions. This is sufficient in one salt, $(TMTSF)_2ClO_4$, for it to be superconducting at atmospheric pressure, the first such example discovered. Further modifications of the TMTSF molecular framework have produced several molecules with supercon-ducting salts. Anion variations affect T_c, though in ways that are not yet well understood. Among these compounds, BEDT-TTF (II) as its $[Cu(NCS)_3^2]^-$ salt gives the highest T_c (10.4 K). [Ni(dmit)$_2$] (IV) is also interesting because it shows that the metal ion does not inhibit superconductivity. It accepts rather than donates electrons to give, for example, [Me$_4$N][Ni(dmit)$_2$] which becomes superconducting at 5 K under 7 kbar pressure.

The T_c values of these compounds are modest compared to the copper oxide superconductors described in section 10.9, but they are none the less interesting, and potentially important. Their requirements for superconductivity appear to be: (i) their packing with weak intermolecular contacts in two directions, suppressing Peierls distortions, and (ii) their ability to gain or lose electrons, producing a part-filled valence or conduction band. It remains unclear how the structural and chemical features control T_c when these requirements are fulfilled.

The development of other 'metallic' properties in organic molecules is also beginning. Stable organic radicals can give interesting magnetic properties, and one example of an organic ferromagnet has been produced. This was formed by polymerising the biradical (VI). The monomer is antiferromagnetic, but it appears that in the polymer, the physical proximity of the unpaired electrons causes a parallel alignment of the spins, though the mechanism for this is unknown.

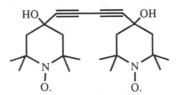

(VI) BIPO — 1,4-bis(2,2,6,6-tetramethyl-4-oxy-4-piperidyl-1-oxyl)butadiin.

11.7   POLARIZED COVALENT BONDING: SILICATES

Few differences appear between polarized and unpolarized covalent bonds, apart from the influence of polarity on chemical reactions. However, a few bond types are distinctive, and this section examines the Si–O bond and silicates. The most important characteristic of this bond is its stability. Most Si–O bonds are inert to hydrolysis or to rearrangement unless the temperature exceeds about 500°C, when interconversion of silicates may occur. The main cause of this stability must be the d_π-p_π donation discussed in section 9.4, but such other factors as the absence of lone pairs on Si probably contribute to the kinetic stability. Here we can only outline the remarkable range of silicates that exist as a consequence. In their way they rival the range of organic compounds resulting from the kinetic stability of C–H, C–C, C–N and C–O bonds.

The characteristic structural unit of the silicates is the SiO$_4$ tetrahedron. Although in the older literature, SiO$_2$ and silicates are often represented as containing Si^{4+} and O^{2-}, the tetrahedral geometry and the relatively high electrone-gativity of silicon indicates that polarized covalent bonding is better than this ionic description. In these tetrahedra, some or all of the oxygen atoms may bridge to other silicon atoms, or may exist as −OH, or −O⁻ with their charges compensated by metal cations. A further crucial feature of the silicate minerals is the ability of Al to replace Si. Charge balance is retained by introducing additional cations (one positive charge per Al). The Al−O bonds are more polar than Si−O bonds, and the greater δ^- charge carried by oxygen leads to changes in reactivity. The substitution of Al for Si leaves the bond character and geometrical preferences otherwise unchanged.

11.7.1 Silicate structures

The ability of the SiO_4 tetrahedra to form links gives rise to the richness of silicate structures and chemistry. However, they can be divided into only five families. The simplest silicates, called *orthosilicates*, contain isolated $[SiO_4]^{4-}$ tetrahedra (Fig. 11.19a); they are not very common, but two are widely distributed: olivine, Mg_2SiO_4, and the semiprecious gemstone zircon, $ZrSiO_4$. Pairs of tetrahedra linked by one Si−O−Si bridge give the $[Si_2O_7]^{6-}$, pyrosilicate, ion (Fig. 11.19b) which

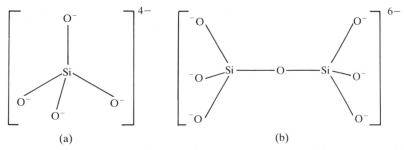

(a)   (b)

Fig. 11.19 — Silicate structures with zero or one Si−O=−Si bridge (a) the isolated SiO_4 tetrahedra of orthosilicates, (b) SiO_4 tetrahedra sharing one corner in pyrosilicates.

generally has a bond angle near 180° consistent with $p_\pi \rightarrow d_\pi$ donation from O to Si (section 9.4.1). It occurs naturally in a few rare minerals.

With two shared corners, more arrangements become possible. The tetrahedra can be linked in rings of various sizes, including $[Si_3O_9]^{6-}$ (Fig. 11.20a) and $[Si_6O_{18}]^{12-}$, found in beryl, $Be_3Al_2Si_3O_9$. The Al^{3+} in the latter are octahedrally

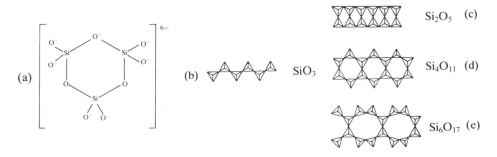

Fig. 11.20 — Silicate structures with two or three Si−O−Si bridges (a) $[Si_3O_9]^{6-}$ rings, (b) single chains (pyroxenes), (c)–(e) varieties of double chains; amphiboles show type (d). Reproduced with permission from Wells, A. F. (1984) *Structural inorganic chemistry*, Oxford University Press, Oxford, Fig. 23.12.

coordinated cations, rather than replacements for tetrahedrally coordinated silicon atoms in the anions. The substitution of as little as 0.3% of the Al^{3+} by Cr^{3+} produces emerald rather than beryl, illustrating one of the sources of diversity in silicates: the combination of different cations with particular silicate anions.

Silicate tetrahedra sharing two vertices also form chains (Fig. 11.20b), while doubled chains result when a proportion of the tetrahedra share three corners (all, half, or a third in Figs 11.20c–e). Their repeating units are respectively $[SiO_3]_n^{2n-}$, $[Si_2O_5]_n^{2n-}$, $[Si_4O_{11}]_n^{5n-}$, $[Si_6O_{17}]_n^{7n-}$. Minerals containing these chains include pyroxenes and amphiboles. In their crystals, the chains are parallel, with the cations between them. As a result, the materials are strong along the length of the chains but split easily between them. Thus they are fibrous minerals and include the different varieties of asbestos, e.g. tremolite (an amphibole), $Ca_2Mg_5(OH)_2[Si_4O_{11}]_2$.

With three shared corners, the SiO_4 units are normally arranged in layers, as in Fig. 11.21a, whose repeating unit has formula $[Si_2O_5]_n^{2n-}$. More complex layer structures and double layers are also formed, when some tetrahedra share all four vertices. All these layer silicates cleave readily between the layers, and they therefore include the micas and talc, $Mg_3(OH)_2[Si_4O_{10}]$ (Fig. 11.21a–b).

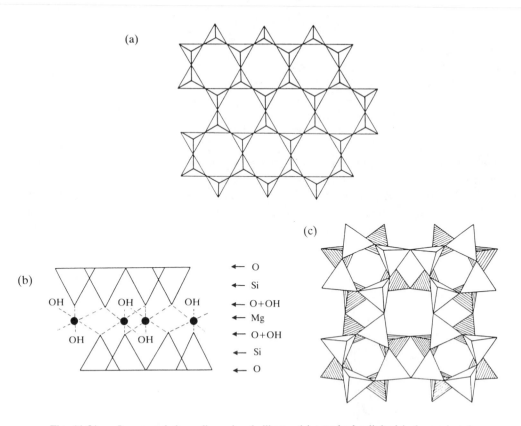

Fig. 11.21 — Layer and three-dimensional silicates (a) tetrahedra linked in layers in talc, $Mg_3(OH)_2Si_4O_{10}$, (b) side view of layers in talc, showing the location of Mg^{2+} and OH^- ions, (c) tetrahedra linked in three dimensions in feldspar. Note the small cavity (contrast Fig. 11.22). Reproduced with permission from Adams, D. M., *Inorganic solids*, 1974, p. 254.

In the final stage, every tetrahedron shares all four corners. This gives silica itself, SiO_2, which exists in a number of forms in depending on the arrangement of the linked tetrahedra. In quartz (stable at room temperature), the tetrahedra form left- or right-handed spirals, giving it the ability to rotate the plane of polarised light. It is converted on heating into tridymite (stable in the range 870°–1470°C) and then cristobalite (stable 1470°–1710°C, the melting point).

11.7.2   Aluminosilicates
This sequence of increasingly interlinked SiO_4 tetrahedra tells only half the story. At each level, replacement of some Si^{4+} by Al^{3+}, produces aluminosilicates. Thus in the layer silicates, the mica muscovite, $[KAl_2(OH)_2]^{5+}[Si_3AlO_{10}]^{5-}$, has 25% Al^{3+} replacing Si in [4] coordinate sites. As is common, it also contains Al^{3+} ions in [6] coordinate sites. Clay minerals are also layer aluminosilicates, with higher ratios of Al^{3+} substitution, and hence more cations. Water often intercalates between the sheets, causing their spacing to increase. This has the result that the clays expand or contract, depending on the conditions. Substitution by Al^{3+} is most important in SiO_2 itself, converting it into a great variety of framework silicates. These include many important minerals, especially the felspars (Fig. 11.21c) such as orthoclase, $K[AlSi_3O_8]$, and anorthite, $Ca[Al_2Si_2O_8]$, with 25% and 50% Al replacement.

11.7.3   Zeolites
The group of framework aluminosilicates that have recently become of greatest chemical interest are the *zeolites*. Their key characteristic is their very open silicate frameworks. These contain cavities which can hold guest molecules. The name zeolite means 'boiling stone', because natural hydrated zeolites give off water on heating. The natural zeolites have been supplemented by a variety of synthetic examples, with a wide range of size and arrangement of cavities. Fig. 11.22 show two typical examples. The cavity in zeolite-A is rather small, with a diameter of 420 pm. In faujasite the cavity is considerably larger, diameter 1200 pm, but its entrance is restricted to a diameter of 740 pm.

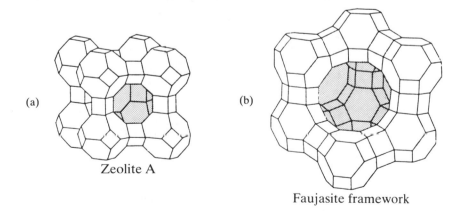

(a)

Zeolite A

(b)

Faujasite framework

Fig. 11.22 — Zeolite frameworks, showing the cavities. An Si or Al atom is located at each node of the framework (O atoms omitted). (a) Zeolite-A, (b) Faujasite. Besed on: Dwyer, J. & Dyer, A. (1984) *Chem. & Ind.* p. 237.

The properties that make zeolites of interest are (a) the ability to exchange the cations contained in the cavities, leading to use as *ion-exchange* materials; (b) the ability for molecules of appropriate size and shape to enter the cavities while others are excluded. This property is used in *molecular sieves*, for example, removing water from organic liquids; (c) catalytic activity associated especially with $-O^-$ groups adjacent to Al^{3+} ions. These centres act as powerful Lewis acid catalysts, for example isomerising hydrocarbons. Al_2O_3 shows the same behaviour, but it reacts indiscriminately with any hydrocarbon. Fig. 11.23 shows three ways in which the structure of a zeolite may control its reactivity. Some potential reactants may be unable to enter the cavity (e.g. in (a), a branched-chain hydrocarbon). Some products may be unable to leave the cavity, e.g. (b) o- and m-xylenes; they will continue isomerising, escaping only as the p- form. In the third mode of control (c), a bulky transition state cannot form in the cavity, limiting the possible products.

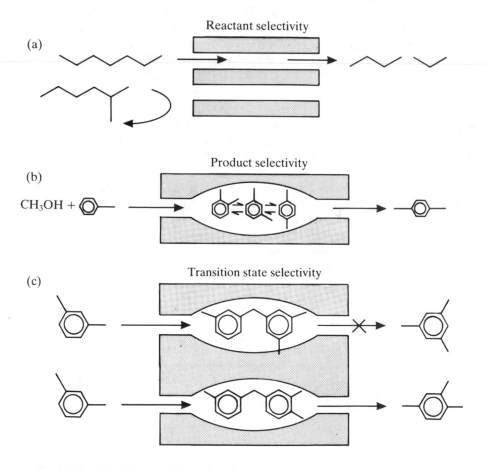

Fig. 11.23 — Catalytic control by zeolites. (a) Reactant selectivity. A linear but not a branched alkane fits in the cavity. (b) Product selectivity. Only p-xylene can escape from the cavity. (c) Transition state selectivity. During methyl group transfer, only the transition state to 1,2,4 substitution fits in the cavity. Reproduced with permission from Csicsery, S. M. (1985) *Chem. in Britain*, p. 473.

11.8 MODIFIED METALLIC BONDING

In the compounds considered in this section highly electropositive metals are combined with with non-metals. On the bond triangle (Fig. 11.1) they are the mirror image of the silicates. They retain some metallic properties, accompanied by ionic *and* covalent structural characteristics. Thus, they are truly intermediate in their character.

11.8.1 Zintl phases

The Zintl phases† combine an element of group 1 or 2 with one from groups 13 to 15. They are extremely numerous, with at least 300 binary compounds identified; ternary compounds can also be formed, though these have been less explored. Crystal structures are known for about half of them, but study of their physical properties is less complete. A most elegant approach has been developed to explain the unusual character of their crystal packing. It involves an application of the *isoelectronic* principle, reaching even further than those already encountered.

Some of the Zintl phases have the expected stoicheiometries for ionic compounds: Na_3As can be formulated as Na^+As^{3-}, and it has isolated As atoms in its structure, though their bonding is presumably strongly polarized. In the majority of these phases, the electronegative (B) atoms are bonded to others of the same type. These B-atoms form a framework whose interstices contain the A-atoms. A typical example is NaTl, in which the Tl atoms have the [4] coordinate diamond structure (Fig. 11.24a; cf. Fig. 11.2). The covalent bonds in diamond require four electrons per atom, but Tl has only three. Each Na is therefore seen as transferring one electron to the Tl framework, and the compound can be written as $n[Na]^+ [Tl]_n^{n-}$. A similar arrangement is found for $CaIn_2$ (i.e. $nCa^{2+} 2[In]_n^{n-}$, though the tetrahedral In network is somewhat distorted. These structures also occur with other group 13 elements, e.g. LiGa, LiIn, $BaTl_2$.

Just as adding one electron to Ga, In, and Tl makes them isoelectronic to diamond, so Si-Pb can become equivalent to P, and P-Sb to Se. Thus, NaSi ($4Na^+$ Si_4^{4-}) contains Si_4^{4-} tetrahedra resembling P_4 (Fig. 11.24b), as does $BaSi_2$ ($2Ba^{2+}$ Si_4^{4-}). In LiP, the phosphorus atoms form spiral chains with the same geometry as elemental Se and Te ($nLi^+ [P_n]^{n-}$).

Some of the anion groupings are not identical to those of the corresponding elements, but are clearly consistent with their apparent valency. Thus, $SrSi_2$ and LiGa contain 3-dimensional networks of three-coordinate Si atoms ($Sr^{2+} 2[Si_n]^{n-}$; $Li^+ [Ge_n]^{n-}$). In $CaSi_2$, however, the $[Si_n]^{n-}$ unit forms a layer with the geometry of elemental As (Fig. 11.24c). Other Zintl phases with more complex formulae contain more than one type of anion unit. Li_7Si_2 corresponds to ($28Li^+ 4Si^{4-} 2Si_2^{6-}$); the Si_2^{6-} ions are isoelectronic with Cl_2.

This summary does not exhaust the structural complexities of the Zintl phases, and they will undoubtedly produce more surprises as further structures are determined. Although the principles underlying the individual structures are well-defined, it is not yet obvious what controls the adoption of one structure rather than

---

† Named after Edward Zintl, who first prepared many of them. An excellent description of them can be found in Schäfer, H. *et al.* (1973).

(a)

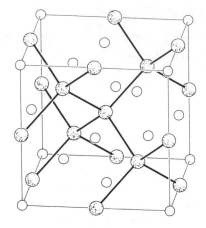

(b)

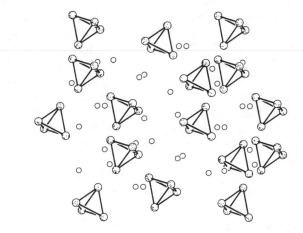

(c)

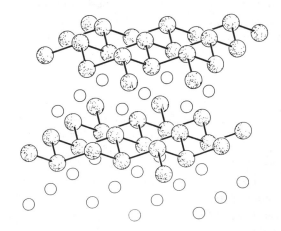

Fig. 11.24 — Zintl phases (a) NaTl, emphasizing the Tl framework, (b) NaSi, showing the Si_4 tetrahedra (c) $CaSi_2$ showing the $(si)_n$ layers. For sources of data, see Schäfer, H. *et al.* (1973).

another, the Si layers of $CaSi_2$ in contrast to the 3-dimensional network of $SrSi_2$, for example. The formation of the anion networks can be rationalized by the transfer of electrons from the cations, giving a combination of covalent and ionic bonding. However, electrical conductivity measurements and other evidence shows that the electron transfer is incomplete, so that the compounds retain metallic properties as well. Perhaps, following Humpty Dumpty, they should be named 'meta–ioni–covalent' compounds.

11.8.2   Cs and Rb sub-oxides

The oxides of the alkali metals should surely be among those compounds showing ideal ionic behaviour. For the most part they do conform to this expectation, though their chemistry is complicated by the existence of stable MO_3, MO_2, and M_2O_2 oxides as well as M_2O (section 4.9.1). However, a very strange group of oxides of Rb and Cs have been discovered which can in no way be described as ionic. They have high M:O ratios and are therefore known as sub-oxides, and they retain much of the character and chemical behaviour of the parent metals. Although the first examples were reported in 1909, they have become well-understood only through the elegant work carried out by A. Simon in the 1970's, overcoming the extreme handling difficulties associated with their fierce reaction with O_2, H_2O, etc. and ready decomposition (sometimes below room temperature) (Simon 1979).

The formulae of the sub-oxides are bizarre, and give little clue to their structures. Typical examples, among the total of about ten, are:

$$Rb_6O, \quad Rb_9O_2, \quad Cs_4O, \quad Cs_7O, \quad Cs_{11}O_3;$$

they also occur with mixed metals, e.g. $Rb_7Cs_{11}O_3$. X-ray crystallography has identified the basic structural unit as an M_6O octahedron (Fig. 11.25a). This is linked in two ways. With Rb, two share one face to give paired octahedra: Rb_9O_2 (Fig. 11.25b), while with Cs, three octahedra share adjacent faces in $Cs_{11}O_3$ (Fig. 11.25c). Assuming standard oxidation states, these units should both be positively charged:

$$Rb_9O_2: \quad +9-4=+5; \quad Cs_{11}O_3: \quad +11-6=+5$$

As the solids contain no anions the electrons removed should be in delocalized metallic orbitals. This is confirmed by the retention of metallic conductivity in the sub-oxides.

These structural units are on their own in two of the oxides and are combined with metal ions in the others:

$$Rb_6O = 3Rb + Rb_9O_2$$
$$Cs_4O = Cs + Cs_{11}O_3$$
$$Cs_7O = 10Cs + Cs_{11}O_3 \text{ (Fig. 11.25d)}$$
$$Rb_7Cs_{11}O_3 = 7Rb + Cs_{11}O_3.$$

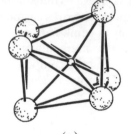

(a)

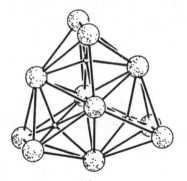

(b)

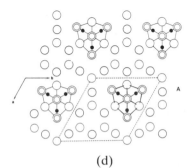

(c)

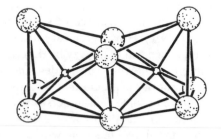

(d)

Fig. 11.25 — Alkali metal sub-oxides (a) M_6O octahedron, (b) Rb_9O_2 paired face-sharing octahedra, (c) $Cs_{11}O_3$ triple face-sharing octahedra, (d) unit cell of Cs_7O, containing $Cs_{11}O_3$ units and Cs atoms. Reproduced with permission (d) from Simon, A. (1979) *Structure and bonding* **36** 81.

Interestingly, the mixed metal oxide retains the $Cs_{11}O_3$ unit, showing that Rb and Cs cannot exchange environments. These more complex species can be seen as alloys of large 'metals', Rb_9O_2 or $Cs_{11}O_3$, and smaller ones, Rb or Cs.

The bond lengths agree with this description. The metal–metal distances within clusters are quite short (Rb–Rb 352 pm; Cs–Cs 380 pm), but between clusters the distances are much larger (Rb–Rb 510 pm; Cs–Cs$_3$527 pm), comparable to those in the metals (485 and 532 pm).

As with the Zintl phases, it is worth emphasizing that we cannot predict *which* phases will be stable, even though we can understand something of their specific structures. Appropriately, therefore, this book ends with two questions: how can we use our knowledge of the bonding in individual compounds to predict what other formulae and structures will be stable, and, how can we identify from the structures, which compounds will have interesting and useful properties? It is in these directions that our knowledge of bonding and structure will surely be extended in the future.

FURTHER READING

Semiconductors
Delaney, C. (1980) *Electronics for the physicist*. Ellis Horwood, Chichester.
Morris, N. M. (1976) *Semiconductor devices*. Macmillan, London.
Tuck, B. & Christopoulos, C. (1986) Physical Electronics. Edward Arnold, London.

Ionicity
Mooser, E. & Pearson, W. B. (1959) *Acta Cryst.* **12** 1015.

Intercalation compounds
Greenwood, N. N. & Earnshaw, A. (1984) *Chemistry of the elements*. Pergamon, Oxford, p. 313f.
Levy, F. (ed.) (1979) *Intercalated layer materials*. Reidel, Dordrecht.

Molecular metals
Cox, P. A. (1987) *The electronic structure and chemistry of solids*, Oxford University Press, Oxford. p. 175f.
Inokuchi, H. (1988) New organic superconductors. *Angew. Chem. Int.* **27** 1747.
Williams, J. M. (1985) Rational design of synthetic metal superconductors. *Prog. Inorg. Chem.* **33** 183.

Silicates
Adams, D. M. (1976) *Inorganic solids*. Wiley, Chichester
Wells, A. F. (1984) *Structural inorganic chemistry*. Oxford University Press, Oxford.

Zeolites
Csicsery, S. M. (1985) *Chem. in Britain*, p. 473.
Dwyer, J. & Dyer, A. (1984) *Chem. & Ind.* p. 237.

Zintl phases and sub-oxides
Phillips, J. C. (1970) *Rev. Modern Phys.* **42** 317.
Schäfer, H., Eisenmann, B. & Müller, W. (1973) Zintl phases: transitions between metallic and ionic bonding. *Angew. Chem. Int.* **12** 694.
Simon, A. (1979) Structure and bonding with alkali metal suboxides. *Structure and Bonding* **36** 81.

Text references
Villars, P. (1983) *J. Less Common Metals* **92** 215.

Appendix: Tables of radii

A.1 CARBON-BASED COVALENT RADII

A.1.1. Single bond radii (pm)

Element	H
Radius†	29.9

Element	Be	B	C	N	O	F
Radius	106	83.0	76.7	70.2	65.9	61.9
Source	b	a	a	a	a	a

Element		Al	Si	P(III)	S(II)	Cl
Radius		118	109.0	108.8	105.2	102.3
Source		b	a	a	a	a

Element		Ga	Ge	As(III)	Se(II)	Br
Radius		125	122	119.6	120.3	119.9
Source		b	b	a	a	a

Element		In	Sn	Sb(III)	Te(II)	I
Radius		141	139	137	139.1	139.5
Source		b	b	b	a	a

†The quoted radius for H applies to the actual position of the H-nucleus, as determined by neutron diffraction. With X-ray diffraction, the observed position of the H-atom is the centre of gravity of its electron cloud, which lies about 10 pm closer to the attached atom. This gives an apparent H-atom radius close to 20 pm.

Sources
a: From tabulations and averages of $C(sp^3)$-X distances in Allen, F. H., Kennard, O., Watson, D. G., Brammer, L., Orpen, A. G., & Taylor, R. (1987) *J. Chem. Soc. Perkin II* p. S1, subtracting 76.7 pm for the radius of carbon.
b: From individual CH_3-X distances, again subtracting 76.7 pm.

A1.2 Multiple bond radii (pm)

Element	C	N	O
Single	76.7	70.2	65.9
Double	66.1	61.8	54.9
Triple	59.1	54.5	

Source: Allen *et al.* (1987), as section A1.1.

A.2   VAN DER WAALS RADII

Element		N	O	F	
Radius		155	152	147	
Element	Si	P	S	Cl	
Radius	210	180	180	175	
Element	Ge	As	Se	Br	
Radius	195	185	190	185	
Element	Sn	Sb	Te	I	Xe[†]
Radius	210	205	206	198	200
Element		Bi			
Radius		215			

[†]This value for Xe seems to be more appropriate for compounds than the value of 216 pm found in the element; Alcock, N. W. (1972) *Adv. Inorg. Chem. Radiochem.* **15** 4.

Source: Bondi, A. (1964) *J. Phys. Chem.* **68** 441.

A.3   METALLIC RADII (12-COORDINATE) (PM)

Element	Radius	Element	Radius	Element	Radius
Ag	144.5	In	166.3	Ru	133.9
Al	143.2	Ir	135.7	Sb	159
Au	144.2	K	237.6	Sc	164.1
Ba	224.3	La	187.7	Sm	180.2
Be	112.8	Li	156.2	Sn	162.3
Bi	170	Lu	173.4	Sr	215.1
Ca	197.4	Mg	160.2	Ta	146.7
Cd	156.8	Mn	126.4	Tb	178.2
Ce	182.5	Mo	140.0	Tc	136.0
Co	125.2	Na	191.1	Th	179.8
Cr	128.2	Nb	146.8	Ti	146.2
Cs	273.1	Nd	182.1	Tl	171.6
Cu	127.8	Ni	124.6	Tm	174.6
Dy	177.3	Os	135.3	U	156
Er	175.7	Pb	175.0	V	134.6
Eu	204.2	Pd	137.6	W	140.8
Fe	127.4	Pm	181.0	Y	180.1
Ga	141.1	Pr	182.8	Yb	174.0
Gd	180.2	Pt	138.7	Zn	139.4
Hf	158.0	Rb	254.6	Zr	160.2
Hg	157.3	Re	137.5		
Ho	176.6	Rh	134.5		

Source: Teatum, E., Gschneidner, K., & Waber, J. (1960) *Compilation of calculated data useful in predicting metallurgical behaviour of the elements in binary alloy systems*, LA-2345, Los Alamos Scientific Laboratory.

A.4   IONIC RADII

A.4.1   Cation radii (6-coordinate) (pm)

Radii are quoted for common oxidation states up to +3 (4 for Hf, Th, Ti, U, and Zr).

Element	Radius	Element	Radius	Element	Radius
Ag(+1)	129	Gd(+3)	107.8	Rh(+3)	80.5
Al(+3)	67.5	Hf(+4)	85	Ru(+3)	82
Au(+1)	151	Hg(+1)	133	Sb(+3)	90
Au(+3)	99	Hg(+2)	116	Sc(+3)	88.5
Ba(+2)	149	Ho(+3)	104.1	Sm(+3)	109.8
Be(+2)	59	In(+3)	94.0	Sr(+2)	132
Bi(+3)	117	Ir(+3)	82	Ta(+3)	86
Ca(+2)	114	K(+1)	152	Tb(+3)	106.3
Cd(+2)	109	La(+3)	117.2	Th(+4)	108
Ce(+3)	115	Li(+1)	90	Ti(+2)	100
Ce(+4)	101	Lu(+3)	100.1	Ti(+3)	81.0
Co(+2)	79/88.5[†]	Mg(+2)	86.0	Ti(+4)	74.5
Co(+3)	68.5/75[†]	Mn(+2)	81/97.0[†]	Tl(+1)	164
Cr(+2)	87/94[†]	Mn(+3)	72/78.5[†]	Tl(+3)	102.5
Cr(+3)	75.5	Mo(+3)	83	Tm(+3)	102.0
Cs(+1)	181	Na(+1)	116	U(+3)	116.5
Cu(+1)	91	Nb(+3)	86	U(+4)	103
Cu(+2)	87	Nd(+3)	112.3	V(+2)	93
Dy(+3)	105.2	Ni(+2)	83.0	V(+3)	78.0
Er(+3)	103.0	Pb(+2)	133	Y(+3)	104.0
Eu(+2)	131	Pd(+2)	100	Yb(+2)	116
Eu(+3)	108.7	Pm(+3)	111	Yb(+3)	100.8
Fe(+2)	75/92.0[†]	Pr(+3)	113	Zn(+2)	88.0
Fe(+3)	69/78.5[†]	Pt(+2)	94	Zr(+4)	86
Ga(+3)	76.0	Rb(+1)	166		

†Low spin and high spin values (section 8.2.3).

Source: Shannon, R. D. (1976). 'Revised effective ionic radii in halides and chalcogenides' *Acta Cryst.* **A32**, 751. This includes further oxidation states and coordination numbers.

A.4.2   Anion radii (6-coordinate) (pm)

Element	Radius	Element	Radius
Cl(−1)	167	O(−2)	126
Br(−1)	182	S(−2)	170
F(−1)	119	Se(−2)	184
I(−1)	206	Te(−2)	207

Source: as A.4.1.

Index